Adam Politzer, George Stone

The Anatomical and Histological Dissection of the Human Ear

In the Normal and Diseased Condition

Adam Politzer, George Stone

The Anatomical and Histological Dissection of the Human Ear
In the Normal and Diseased Condition

ISBN/EAN: 9783337255855

Printed in Europe, USA, Canada, Australia, Japan

Cover: Foto ©berggeist007 / pixelio.de

More available books at **www.hansebooks.com**

THE

ANATOMICAL AND HISTOLOGICAL DISSECTION

OF

THE HUMAN EAR

IN THE NORMAL AND DISEASED CONDITION.

BY

DR. ADAM POLITZER,

PROFESSOR OF OTOLOGY IN THE IMPERIAL ROYAL UNIVERSITY OF VIENNA, ETC.

Translated from the German

BY

GEORGE STONE.

WITH 164 ILLUSTRATIONS AND 1 PLATE IN THE TEXT.

LONDON:

BAILLIÈRE, TINDALL AND COX,

20 & 21, KING WILLIAM STREET, STRAND.

1892.

PREFACE TO THE TRANSLATION.

A GLANCE at the original work convinced me that I had to deal with a task different from that which falls to the lot of one who undertakes to render into readable English an ordinary handbook of Otology, where considerable latitude of style is allowed.

In the present work, where precise rules as to special dissections and modes of preparing the various parts of the organ of hearing are given, it has frequently been unavoidable to translate almost literally.

I hope, however, that reasonable limits have not been exceeded when it was found advisable to sacrifice smoothness of style to faithful rendering of very minute, though most important details, in the exhaustive treatment of the subject by so eminent an authority as Professor Politzer.

Attention is called to the lettering in 29 of the 164 illustrations, which were printed from clichés used for the English edition of the Author's Text-Book of Diseases of the Ear.

GEORGE STONE.

88, RODNEY STREET, LIVERPOOL,
October, 1891.

To Mr. George Stone, Aural Surgeon, Liverpool.

DEAR MR. STONE,

I have read with much interest the translation of my work, "Zergliederung des Gehörorgans," which you were good enough to send me; and I cannot but congratulate you on the patience bestowed upon it, and on the remarkable exactness in the rendering, both as regards scientific interpretation, and clearness and conciseness of style, which invest the translation with the character of an original work.

As our *confrères* in England and America, in common with the Otologists in other parts of the world, highly esteem the memory of your great countryman, Joseph Toynbee, the founder of the Pathological Anatomy of the Ear, and have always shown the liveliest interest in matters of exact anatomical science, they will, no doubt, give your work a favourable reception.

Accept my thanks for having honoured my monograph with a translation into your beautiful mother-tongue, and be assured of my best wishes for the success of your conscientious labours.

Yours sincerely,

PROF. DR. A. POLITZER.

VIENNA, *October*, 1891.

CONTENTS.

PAGE

INTRODUCTION.

THE progress which during recent years has been attendant upon medical research in the field of Otology has been most marked. No doubt the sound pathological methods pursued in modern investigation into this subject have contributed very largely to this desirable condition of things, and have materially assisted to popularize the hitherto neglected study of the anatomy of the human ear. Ever since the labours of Toynbee, it has been fully realized that a knowledge of the pathological anatomy of the organ under discussion is absolutely necessary to a right comprehension and successful treatment of its diseases. The complicated character of the structure of the organ of hearing, and the involved nature of its connections with the adjoining structures and cavities, are such that they can only be intelligently studied by frequent practice in the preparation of specimens. Those who look for instruction on the subject of the structure of the ear from preparations ready to hand, or are satisfied with pictorial illustrations and letter-press description, will never acquire a thorough knowledge of the topographical position of the various parts of the ear. To the aural surgeon in particular, practice in dissection work is indispensable. It is only by unremitting study and manipulation of specimens that he can hope to attain to that degree of confidence so necessary to produce satisfactory results, when performing an operation in such close proximity to organs of vital importance. If any further argument were necessary to show that a methodical study of models is a *sine quâ non* of proficiency in this department of medical research, it is to be found in the fact that it is only by a thorough acquaintance with the topographical relations of the parts of the normal ear that anything like a safe theory can be formulated as to the conditions obtaining in the abnormal ear under observation. The beautiful anatomical work of Scarpa, Breschet, and the artistic preparations of Sömmering, Ilg, Hyrtl, and others, are greatly to be admired, although the methods of dissection bequeathed them by the older anatomists could only be denominated imperfect. For ordinary purposes, however, a more simple type of model suffices.

Specimens should be prepared either for practice in dissection, or for illustration in teaching; and the simpler their construction, provided of course they are scientifically true, the more efficacious will they prove.

The first recorded attempt to arrive at a knowledge of the organ of hearing by the assistance of dissection, dates back to the sixteenth and seventeenth centuries, when the revival of the science of anatomy gave so great an impetus to the study of such phases of medical science. In spite of this, however, Fallopius, Ph. Ingrassias, Bartholomeo Eustachio, Valsalva, Cassebohm, Casserius, Fabricius ab Aquapendente, have transmitted to posterity but few practical hints relative to the dissection of the human ear. From this period, down to a comparatively recent date, the subject has hardly attracted that attention and consideration which its importance merits. Toynbee, Hyrtl, Van den Broek, Sappey, Von Tröltsch, Kölliker, Voltolini, Moos, Zaufal, Brunner, Rüdinger, Zuckerkandl, Lucæ, Wendt, Schalle, Schwalbe and others, while they have given to the world an elaborate wealth of detail relative to aural anatomy, have hardly treated the matter with that comprehensive scope which is so necessary to enable students to obtain a grasp of the whole subject.

This fact, combined with the expressed wish of many of those who attended my lectures, has induced me to venture to bring together in a clear manner the result of modern scientific research in this most interesting field of anatomical inquiry. For the past twenty-five years I have pursued, during the intervals of leisure from my professional otological work, an uninterrupted study of the normal and pathological anatomy and histology of the ear, and I may say that the resultant knowledge and experience thus gained, together with matter collected for lectures, form a basis which has proved invaluable to me in preparing the present work.

In dealing with the various methods of preparation which are employed in the anatomical and histological dissection of the ear, I have endeavoured to be as minute as possible, and the aim kept in view has been the finding and the representing with facility the various parts of the organ in their respective positions. The same section of the work treats of the best methods of preserving a collection of preparations, whether wet or dry, and the means to be employed in adjusting and mounting them. I have contented myself, relative to making the technical proceeding available for pathologico-anatomical study, merely with the main features, because the manifold morbid changes in the middle ear, and the frequently complicated alterations in the temporal bone, would render it impossible to give a description of all cases occurring in dissection. I am convinced that the best means of hitting upon the right method of preparation in complicated pathological cases, is to be thoroughly acquainted with the technicalities of the work of dissecting the normal ear.

A considerable portion of the work is devoted to the histological examination of the organ of hearing. I need hardly point out how important is this section of the subject. Those who have studied the results of recent researches in this branch of the subject, and more especially in the pathological histology of the labyrinth, must admit that this phase of the science is yet in its infancy, and that it offers a most enticing field for original investigation. The difficulties, however, presented by the complicated relative structures of the membranous labyrinth are necessarily very great, but these will be reduced in proportion as the methods of examination employed increase in efficiency. As an example of what recent investigation has achieved, I need only allude to the important progress made in the histological representation of the human labyrinth on the basis of the latest results of microscopical investigation, results which undoubtedly are the foreshadowing of great advance in the pathologico-anatomical examination of the labyrinth.

ERRATA

Page 7, line 13 from foot, *for* 'section' *read* 'dissection.'

Page 18, line 9 from top, omit 'that.'

Page 28, line 20 from foot, *for* 'canal. tensor.-tymp.' *read* 'canal. tensor tymp.'

Page 64, line 6 from top, *for* 'mastoid eum' *read* 'mastoideum.'

Page 97, line 16 from foot, *for* 'making' *read* 'mounting.'

Page 105, first line, *for* 'Representation of' *read* 'Making.'

Pages 129, 131, 135, 169, 170 and 176, *for* 'aquæducts' *read* 'aqueducts.'

Page 134, lines 2, 6, 12 from top, and line 17 from foot
Page 135, line 4 from top, and line 18 from foot   } *for* 'aquæduct' *read* 'aqueduct.'
Page 177, line 5 from top

Page 189, line 11 from top, place bracket after the word 'high.'

Page 203, line 11 from top, insert bracket after 'Corti's organ.'

Page 211, line 7 from top, insert 'and' after elder-pith.

Page 232, line 6 from top, *for* 'will be' *read* 'are.'

Page 242, at the lowest extremity of Fig. 152, place the letter 'h.'

In the chapter on 'Preparatory Methods,' *for* 'Gilding' *read* 'Gold-staining.'

THE ANATOMICAL AND PATHOLOGICO-ANATOMICAL DISSECTION OF THE ORGAN OF HEARING.

I.

INSTRUMENTS NECESSARY FOR THE PREPARATION OF THE ORGAN OF HEARING.

THE work of preparing the organ of hearing is materially facilitated by the use of instruments adapted to the purpose. Their number and selection must always depend upon the object of the examination. Those who merely wish to study the coarser anatomical relations of the organ will find a few, such as scalpel, saw, hammer, chisel, fixing

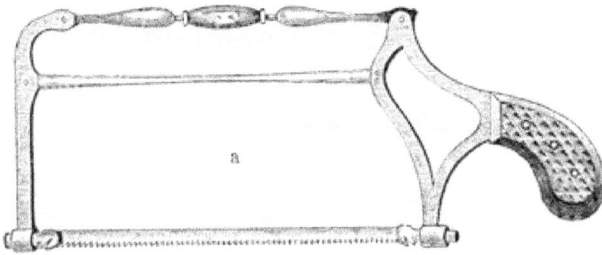

FIG. 1.

forceps, quite sufficient. But those who desire to thoroughly investigate the organ, or to make a complete pathologico-anatomical dissection of it, must employ a more complete set of instruments, and I will here briefly describe what I use in my preparations:

An anatomical saw, similar to an amputation saw, for opening the

FIG. 2.

cranial cavity, and for removing the organ from the base of the skull. For the latter purpose a curved saw (Fig. 1) is recommended, the blade of which can be tightened by means of a steel screw.

A short fret-saw (Fig. 2) for sawing the organ of hearing and the

1

structures of the naso-pharynx out of the base of the skull. The
length of this saw measures 10 cm., the breadth at the posterior
end 1 cm., at the anterior end 3 mm.

FIG. 3.   FIG. 4.

A strong chisel 2½-3 cm. broad for the purpose of chiselling off the
preparation from the base of the skull; further, two smaller straight
ones (Fig. 3), 5 mm. and 3 mm. broad, and a hollow chisel (Fig. 4),

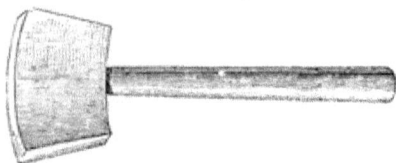

FIG. 5.

3 mm. broad, for removing the walls of the auditory meatus, the
roof of the tympanic cavity, and for opening the mastoid process
and the cavity of the labyrinth.

FIG. 6.

A large and a small iron hammer, also a wooden mallet (Fig. 5);
the former for chiselling out the preparation from the base of the
skull, the latter for the finer work on the bone.

FIG. 7.

An elevator (Fig. 6) and a fixing forceps (Fig. 7) for lifting out from
the base of the skull the preparation when separated with saw and
chisel, but still connected with the soft structures.

FIG. 8.

A large scalpel, so-called cartilage knife (Fig. 8), for separating the
soft parts, when taking out the preparation isolated by saw and chisel,
and for dividing the temporo-maxillary articulations.

Several scalpels of various sizes, for the anatomical preparation of the soft parts, also a narrow knife (Fig. 9) for cutting through the tendon of the tensor tympani and the incudo-stapedial articulation.

FIG. 9.

A raspatory made of hard steel, running to a point and provided with a wooden handle (Burckhardt-Merian), for thorough removal of the soft

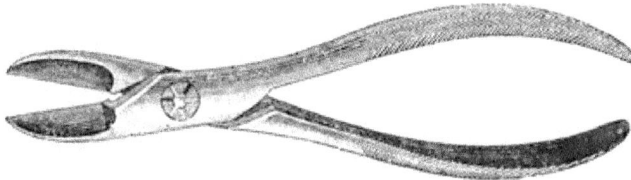

FIG. 10.

structures, the insertions of tendons, and the periosteum from the bone. A good-sized spring bone forceps ending in a point (Fig. 10), for

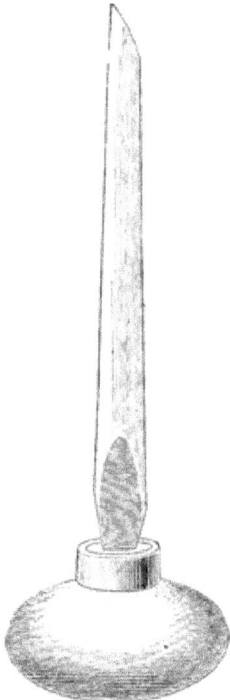

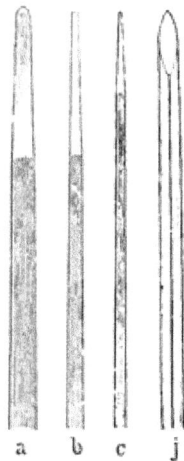

a   b c   j   s   s'

FIG. 11.   FIG. 12.   FIG. 13.

breaking off the larger pieces of bone, and two smaller forceps, for taking away the smaller particles (Leiter, Vienna).

A large and a small curved fret-saw, for making sections and for cutting away the larger portions of bone. For coarser sawing, blades Nos. 2, 3, 4, and for finer, Nos. 00, 0, 1, are used (Schadelbauer, Vienna).

Three or four graving tools (French burin) of different widths (Fig. 11) (Vautier 45-38, Fig. 12, a, b, c), and several adjusting tools

Fig. 14.

(Fig. 12, j) which are chiefly used in the preparation of the osseous labyrinth. For the finer work on the latter, use may be advantageously made of various scrapers, as shown in Fig. 13, s, s'.

A strong, coarse file, two sharp flat files, two or three round files of different strengths, several pointed files and needle files.

Fig. 15.

A drill (Fig. 14).

Several preparing needles, blunt and pointed hooks (Fig. 15). A straight, strong pair of scissors, one blade blunt and the other pointed; a pair of bone scissors; a pair of scissors curved on the flat, and two pairs of finer pointed scissors; a grooved director (Fig. 16), metal and

Fig. 16.

whalebone probes. A strong, but finely-pointed, anatomical pincette (Fig. 17), several smaller, finer pincettes, and various camel's-hair brushes.

A ball syringe or wash bottle; several glass vessels of various sizes, some deep and some shallow; small saucers and watch glasses; conical

Fig. 17.

metal canulas and wooden olives fixed on india-rubber tubing, the latter for condensing and rarefying the air in the external meatus.

A lens or a microscope, for magnifying the details during the preparation, and a reflector for illuminating from time to time the external meatus, the membrana tympani, and the tympanic cavity.

Several glass tubes 2-4 mm. wide, terminating in a point; thin small manometric tubes 5-6 mm. long and 0, 5 mm. wide, for measuring the labyrinthine pressure and for proving the mobility of the stapes, and a number of glass threads, for testing that of the ossicula.

For the preparation of the organ of hearing a small vice is indispensable, notwithstanding the contrary assertions of some anatomists. Although for many manipulations, holding fast the preparation with the left hand in a linen cloth, or resting it against a board, on which two strong projecting ledges are fastened in one corner, is sufficient, yet for cutting fine sections and for carefully chipping off and filing away remains of bone, the fixing of the preparation in the vice cannot be dispensed with. The manner of fixing, however, requires care, because some parts of the temporal bone, especially the outer wall of the tympanic cavity and the parts surrounding the sulcus tymp., may be fractured, and the membrana tympani torn, from too strong pressure of the vice. Best adapted for fixing the preparation in the vice are the squamous portion of the temporal bone, the mastoid process and its place of union with the posterior end of the petrous bone, and for some cases, the anterior portion of the pyramid.

Precautions to be taken against Septic Poisoning while dissecting the Organ of Hearing.

In the preparation of the organ of hearing, the same rules as regards preventive measures against any wounds and their consequences are generally to be observed, as in making post-mortems. The most frequent injuries occur with beginners, chiefly in consequence of faulty handling of the pointed forceps. When the preparation is held in the left hand, it very easily happens that in breaking off particles of the bone, or in opening the upper wall of the tympanic cavity, the forceps slip, and the points forcibly enter the left hand. One should therefore, from the first, accustom one's self so to hold the preparation and forceps that, when the instrument slips, the points may glide in a direction opposite to the palm of the hand. When using the saw in cutting sections, or when removing portions of bone, the preparation should always be fixed in the vice, and never held with the left hand, because any shifting of the saw might cause very dangerous lacerated wounds.

Another frequent cause of injury in the work of preparing lies in the spiculæ of bone which remain behind, when the preparation is taken out of the skull, and which, for want of sufficient care on the part of the operator, cause rather severe scratches on the skin. One should therefore never neglect, before working at a preparation of the temporal bone, to remove all projecting bits of bone with forceps and file.

Bearing in mind the sad examples of cadaveric infection, the utmost caution should be observed, as frequently preparations are made, where the patient has died of some infectious disease, such as typhus, scarlatina and diphtheria. Especially dangerous are cases where individuals have died of sinus phlebitis and pyæmia, and where consequently the greatest caution should be exercised in handling the preparations of the ear.

One ought never to neglect, in such an examination, to see that hands and fingers are free from scratches or excoriations. Where this is the case, it is advisable to postpone the work of preparation, and for the purpose of preserving the object, it should be placed either in Rüdinger's Preservative Fluid (consisting of 800 parts glycerine, 190 parts carbolic acid, and 12 parts alcohol), or in Wywodzew's Solution (thymol 5·0, alcohol 45·0, glycerine 2160·0, aqua destillata 1080·0), and not to undertake the dissection until the scratch in the hand has cicatrized. The same precaution should be observed in the dissection of less dangerous objects, yet one may, in pathological preparations, where the morbid appearances would be too much changed by a preservative fluid, and a postponement of the dissection would be detrimental, so protect the excoriated parts with a bandage, prepared with iodoform-, carbolized-, or sublimate-gauze, or by means of an indiarubber glove, that infection of the injured parts becomes impossible.

As it is well known that there are individuals who have a great disposition to become infected with septic matter, so much so that even with intact skin the poison penetrates through the epidermis into the living tissue and produces sometimes lighter, sometimes severer, forms of post-mortem boils or lymphangitis, it is advisable for individuals so predisposed (Süchtige Individuen, Virchow) to undertake the dissection only of such preparations as have been kept for some days in one of the above-mentioned preservative fluids, and thus been completely disinfected, so that septic poisoning of injured parts is hardly possible.

If, during the dissection, an injury occur to the finger or hand, no matter whether the preparation be disinfected or not, it is advisable, as Nussbaum suggests, to immediately firmly tie the place above the injury with a tape always kept at hand, so as to prevent the reflux of the venous blood and the septic poison from being drawn into the general circulation. After that, a continuous stream of water should be allowed to flow over the wound for several minutes, while, with the other hand, the blood is pressed from the parts above the wound towards it, whereby the poison, on reaching it, is washed out with all the more certainty.

When the wound has been sufficiently washed, the ligature which has been checking the circulation may be removed, and an antiseptic bandage applied. As the effect is the more certain the more quickly the wound is disinfected, it is strongly to be advised that the operator should always have close at hand in a box with a tightly-fitting lid, a small phial containing 5% of carbolized water, a small box with iodoform powder, a small quantity of chloroform, some gutta-percha tissue, Billroth cambric, Bruns' wadding and gauze. The wound, after being irrigated, is washed out with a tuft of cotton-wool dipped in carbolized water, then sprinkled with iodoform powder, wrapped up with a moist piece of gauze, and this either enveloped in Billroth cambric and tied with a tape, or the gauze covering is wrapped round with gutta-percha tissue, which may be fastened by being brushed over with chloroform. The injury heals in a short time by first intention without further consequences.

II.

THE REMOVAL OF THE ORGAN OF HEARING FROM THE DEAD BODY.

THE opening of the cranial cavity, for the purpose of removing the organs of hearing, does not differ from the method practised in the anatomical institutions. This being familiar to every student, a detailed description of the proceeding may here be dispensed with. It should, however, be specially pointed out that in private post-mortems, where, after the corpse has been put upon the bier, all traces of the section should, as far as possible, be concealed, the opening of the cranial cavity must be made with special care, and that the incision from one ear to the other for the separation of the scalp from the calvarium should be made, not across the vertex, but as it is usually done in the institutions of Vienna, across the occipital region. As this is most efficiently carried out by trained, practical, anatomical attendants, it is always advisable, where possible, to secure in private post-mortems such assistance.

The mode of proceeding in the removal of the organs of hearing from the skull differs materially when the body, as in the Anatomical Schools, is at the disposal of the operator, from that of those cases where, having regard to the placing of the body on the bier, every outwardly visible trace of injury to the head must be avoided. Again,

the method will be different in post-mortems where, on account of
co-existing disease at the naso-pharynx and Eustachian tubes, these
parts have also to be examined, from what it will be in those cases
where, during life, disease of these parts could be excluded, so that
their examination would seem to be superfluous.

1. REMOVAL OF THE ORGAN OF HEARING FROM THE DEAD BODY FOR ANATOMICAL PURPOSES.

The simplest proceeding in the removal of the organs of hearing from
bodies intended for anatomical examination, and not requiring special
regard to be taken, but which, under certain circumstances, may also
be observed with bodies about to be laid out, is the following:

After removing the skull-cap in the regular way, the brain is taken
out, and at the same time the acoustic and the facial nerves are
carefully divided close to their exit from the medulla oblongata.
Next, with a strong scalpel, the skin is divided by a perpendicular
incision 8-10 cm. long, commencing 2 cm. behind the insertion of the
auricle, that is, immediately behind the mastoid process; and the
anterior flap of skin, together with the auricle after detachment from
the cartilaginous meatus, is dissected away up to the zygomatic arch.
Now, in case only one of the organs is to be removed, a cut is made
with the anatomical saw, commencing behind the mastoid process
(Fig. 18p, p. 12) and dividing the posterior cranial fossa up to that
point of the medial line of the clivus, which lies midway between the
sella turcica and the anterior edge of the occipital foramen (o). With
a second cut, almost parallel to the transverse diameter of the skull,
the middle cranial fossa is sawn through in a vertical plane, which
unites the middle of the zygomatic arch with the tubercle of the sella
turcica (z s), and separates the most anterior portion of the pars
squamosa, as well as the great wing of the sphenoid and the proc.
pterygoid. close to the superior maxilla. The median ends of these
two cuts (s o) are united by breaking through, with a broad chisel,
the base of the skull in the median line of the sella turcica and the
clivus.

If the base of the skull has been successfully divided, the preparation
can be easily removed. Where this is not the case, the bridges of
bone, remaining behind, must be cut away with the chisel. The preparation is now merely left in connection with the soft structures and
the maxillary joint. In order to disconnect it from the latter, the
muscular and tendinous attachments to the mastoid process must first
of all be quickly dissected up; the preparation should then be pressed
forward and upward, and after cutting through the posterior capsular

wall, the maxillary joint is disarticulated with the cartilage knife. It now suffices to cut through the soft parts of the neck and naso-pharynx, still connected with the preparation, in order to allow the temporal bone to be removed. In such preparations the entire Eustachian tube, with the adjoining parts of the naso-pharynx, are retained.

In a like manner one proceeds in the removal of both organs from bodies which are to be laid out, with this difference, that the cuts with the saw through the base of the skull are not made up to the median line, but only as far as the lateral borders of the sella turcica and the clivus, and that the cuts are joined, on either side, by a chisel breach (Fig. 18 i i′ and u u′), running parallel with the long diameter of the skull. By this means the sella turcica and the clivus will be preserved as a firm bridge between the anterior and posterior parts of the skull, and its collapse will be thus prevented when the body is laid out. If the dissection be made with some care, the cranial cavity filled up with chopped straw or grass, and after accurate readjustment of the calvaria, the cutaneous incision behind the ear and across the occiput be well stitched up, and the body skilfully laid out, nothing will be noticed which might betray any sign of the skull having been interfered with.

The removal of both organs in connection, for definite anatomical purposes, e.g., for showing the angle of inclination of the tympanic membrane, can only be effected when the body is entirely at the disposal of the operator. In this case the anterior saw-cut, made parallel with the transverse diameter of the skull, passes through the entire base of the latter near the tubercul. sellæ (Fig. 18 z z′), while the posterior cut, parallel with the anterior, is also carried through the entire base of the skull behind the mastoid processes and through the middle of the foramen occipitale magnum (m m′). As in the former method so here, after disarticulating the maxillary joint, dividing the soft parts, and loosening the occipito-atloid connection, the preparation is removed. It is a matter of course that in normal anatomical preparations, where auricle and cartilaginous meatus are included, these parts should be allowed to remain in connection with the temporal bone.

2. REMOVAL OF THE ORGANS OF HEARING FROM BODIES WITHOUT EXTERNALLY VISIBLE INJURY TO THE SKULL.

The proceeding differs materially in the removal of the organ from a body where, externally, no trace of the operation be visible. As in these cases the main object is the examination of pathological changes in the organ, the mode of removal will vary according to the diagnosed seat of the disease during life. Thus if, during life, no morbid changes in the external meatus, in the Eustachian tube, and in the naso-pharynx were proved to exist, these portions of the organ will be excluded, and attention be directed mainly to the removal of the tympanic and labyrinthine structures.

(a) Removal of the Organs of Hearing, together with the Naso-pharynx and Eustachian Tubes.

1. A method of removing the organs of hearing, in connection with the naso-pharynx, without apparent external injury to the skull, was first pointed out by Wendt. He made use of the chisel for gouging out the circumscribed parts of the base of the skull, a proceeding which, even with considerable practice, may easily lead to splintering of the external meatus and tympanic cavity. Much safer, though more troublesome, is the method suggested by Schalle,* which we here briefly describe.

After taking off the skull-cap and removing the brain, an incision is first made across both clavicles, from the top of one shoulder to the other, then one on either side, which, commencing 4 cm. behind the auricle, passes over the sterno-cleido-mastoid muscle downward and somewhat forward, and meets the incision made along the clavicle at an acute angle.

Grasping the flap of skin at the upper point of departure, it is separated by a few sweeping strokes, taking as little of the soft parts as possible ; the cartilage of the external meatus is cut through close to the bone, and the maxillary joint, as well as the entire lower edge of the ramus of the corresponding inferior maxilla, are laid bare. When this has been done on both sides, the skin on the anterior surface of the neck is severed from the soft parts, until the lower edge of the maxilla is exposed all round.

The knife is then pushed in a direction corresponding with the origin of the genio-hyoid muscle through the floor of the mouth, and, while keeping close to the lower jaw, the incisions on both sides are enlarged up to the angles of the maxilla ; the tongue is drawn down and cut away with the soft parts of the pharynx down to the lower border of the larynx. Tongue, larynx, trachea, and œsophagus are, for the purpose of gaining space, dissected off from the vertebral column with a few ample strokes and pushed down, or entirely taken out.

After completely removing the soft parts from the cavity of the mouth and naso-pharynx, the capsule of the maxillary joint on both

* Virchow's Archiv, Bd. 71.

sides is divided close to the lower jaw, and the latter drawn so far forward out of the socket that it luxates on the tubercul. articul. In this position the inferior maxilla is fixed by means of a proper chin-holder, and the posterior wall of the pharynx dissected away from the vertebral column.

In order now to saw out the organs of hearing, together with the structures of the nasal cavity and pharyngeal space, the two surfaces of the maxillary joint are broken through with the chisel from below upward, so as to cause, at the corresponding parts of the middle cranial fossa, a gap large enough to allow a narrow saw-blade to be inserted. This having been done in the breach on the left side, the ends of the blade are connected with the bow of the saw in such a way that its handle rests on the side of the cranial cavity. The operator may now proceed with his work in a sitting position, having the head of the body on a level with his shoulder. The cuts should be carried in a direction as nearly as possible corresponding to the long axis of the body, and vertically to the base of the skull. The course which the saw should take is, according to Schalle's description, as follows :

The line of the cut runs from the left breach forward and a little outward, including the anterior median corner of the squamous portion of the temporal bone, then through the great wing of the sphenoid bone, meeting about 1 cm. outward from the superior orbital fissure the roof of the orbital cavity, whence it passes in a curve forward, thus including the greater portion of the latter, reaches, about 15 mm. from the crista galli laterally, the ascending portion of the frontal bone, turns, while keeping close to this in front of the crista, through the foramen coecum to the other side, and meets the right breach in the reverse order. Here it turns in a somewhat short curve outward, and reaches, through the plane of the maxillary joint, without touching the fissura petro-squam., the wall of the cranium again ; keeping close to this, it comes to the posterior edge of the sigmoid sinus, in which it descends with a slight curve to the foramen jugulare, turns, corresponding to the opening of the inferior petrosal sinus, round the tubercul. jugul. dextr., divides the pars basilaris, and avoiding the tubercul. sinistr., passes to the left foramen jugulare ; from here it encircles the left temporal bone in the same manner as on the right side.

But this method, by which both organs of hearing, together with the structures of the naso-pharynx, may be completely removed, proves in practice too complicated and occupies too much time, nor is the object of effacing outward traces of interference quite attained.

2. In a more simple manner it may be done by the following method, which I myself use.

By means of a strong drill (Fig. 14) two perpendicular channels (a a') are made in the anterior cranial fossa, 1 cm. right and left of the crista galli, running downwards through the nasal cavity to the lower surface of the hard palate; through the right channel the small saw represented in Fig. 2 is introduced from the cranial cavity. A cut (a b) is made with this through the base of the skull, passing first in the anterior cranial fossa backwards and somewhat outwards as far as the inner third of the crista sphenoidalis of the lesser wing of the sphenoid

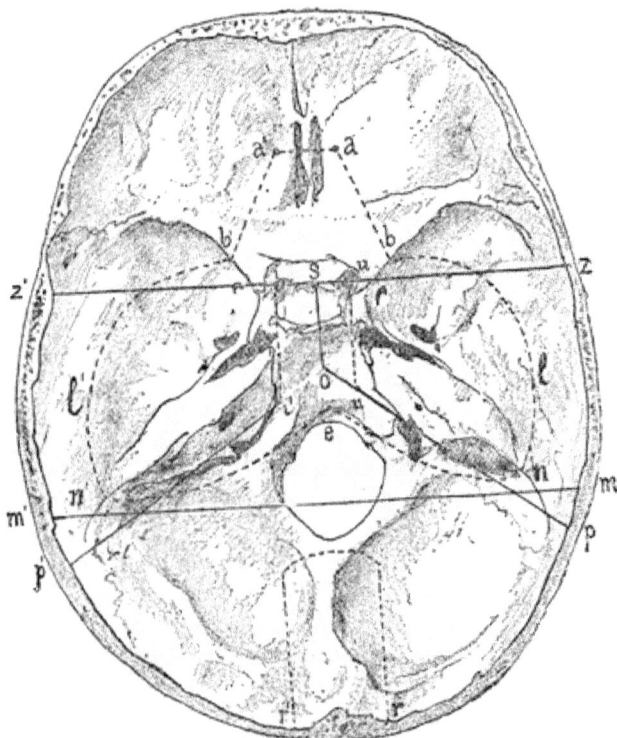

Fig. 18.

bone (2½ cm. from the middle line of the base of the skull). At this point the cut takes a strong sweep outwards in the middle cranial fossa (b l) towards that part of the base of the skull which corresponds to the root of the zygomatic arch, dividing thus the greater wing of the sphenoid bone, and in the temporal bone the maxillary joint and the outer section of the maxillary fossa. When the saw has arrived behind the maxillary joint, that is, in front of the anterior wall of the osseous meatus, a point which projects inwards corresponding to the middle of the clivus of Blumenbach, the instrument should keep close to the point where the upper surface of the pyramid turns to the squamous portion of the temporal bone (l), and the cut be carried as far as the sinus transversus (l n).

This part of the cut divides the middle portion of the osseous meatus and the inner portion of the mastoid cells. The saw now turns from n inwards and somewhat forwards, passes through the open sinus transversus to the foramen jugulare, and from here through the posterior third of the dorsum sellæ, so that the cut falls immediately in front of the condyloid process (e). In the same manner the base of the skull is sawn through symmetrically on the left side also, and the two channels made with the drill are united laterally from the crista galli by a transverse cut. The cutting through the maxillary joint and osseous meatus up to the sinus transversus occupies considerable time, owing to the thickness and density of the bone to be sawn through, while the division of the anterior, and in a great measure also of the middle and posterior cranial fossæ, is accomplished much more quickly.

The piece of the base of the skull round which the saw-cut has been carried is now loosened by a few vigorous blows with the chisel; the firm connections still existing at the posterior and lateral parts of the breach are divided by means of a chisel 4-5 cm. broad, while in the region of the naso-pharynx the separation of the remaining bridges must be effected with the fret-saw, partly from the cranial cavity and partly from the mouth. With the object of detaching the preparation from the soft parts, two parallel cuts (r, r') are carried 3 cm. right and left of the crista occipitalis int. on the occipital bone downwards as far as the posterior edge of the foramen occipitale, and the two cuts united below by means of a curve. Thereby a longish breach is caused on the occiput, which renders it possible to manipulate in a horizontal direction towards the lower surface of the base of the skull. While an assistant, with a long branched forceps (Fig. 7) lays hold of the preparation at the sella turcica and at the posterior cut of the dorsum sellæ, and raises it up towards the front, first the posterior membraneous wall, then the lateral walls of the pharynx, are cut through with a strong scalpel; after that the posterior capsular wall of the maxillary joint is divided on both sides, the joints themselves are disarticulated, and the still existing muscular and ligamentous connections are detached with knife and scissors, to completely separate the preparation. In order to effectually close the gap in the occipital bone when laying out the body, the cuts right and left of the crista occipit. int. are so made that their surfaces are directed inwards, in order that the piece of bone, in being fitted in, may be retained in its position by the mass filling up the cranial cavity.

(b) *Removal of the Organs of Hearing without the Structures of the Naso-pharynx and without the Eustachian Tubes.*

1. Where, from the objective examination during life, no morbid changes in the naso-pharynx and the Eustachian tubes were proved to exist, and where the examination of these parts may thus be dispensed with at the autopsy, the organ can be removed from the

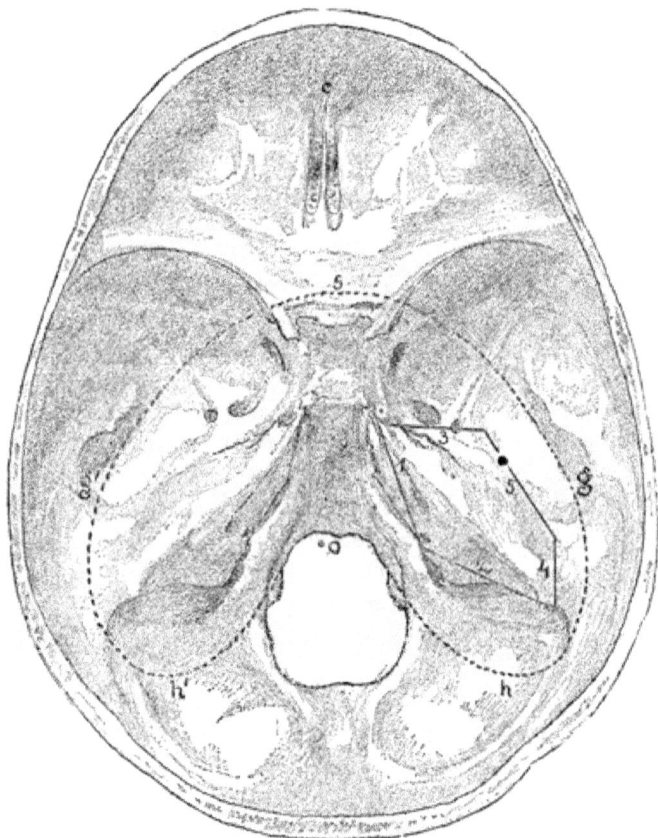

FIG. 19.

skull in a short time, and in a very simple manner. The proceeding is indicated by the lines 1, 2, 3, 4, 5, in Fig. 19. With the broadest chisel (3 cm.), which must be applied as far as possible horizontally and in a direction parallel with the base of the skull, the fissura petro-basilaris is first cut, from the apex of the pyramid to the foramen jugulare (1). As the edge of the chisel advances outwards along the lower surface of the pyramid, the connections of the soft parts of the

temporal bone are also cut below. Then the sinus transversus is
broken through by one or two blows with the chisel, from the foramen
jugulare to where the sin. petros. super. passes into the sin. transvers.
(2) A third perpendicular cut with the chisel (3) is directed outwards
and backwards from the anterior point of the pyramid, between the
latter and the foramen ovale, 1 cm. beyond the foramen spinosum.

The fourth breach (4) made with the chisel to the length of 1 cm.
at the posterior external angle of the middle cranial fossa is joined
with breach 2. The next following cut (5), carried with the chisel as
far as possible outwards at the boundary between the lower surface
of the middle cranial fossa and the vertical squamous portion of the
temporal bone, divides the osseous meatus about the middle. If the
chiselling be not done with care, not only the meatus will be splintered,
but cracks in the bone may be caused in all directions, whereby the
tympanic membrane will be torn and the ossicles dislocated.

Before proceeding to this—the most important part of the operation
—the breaches 1, 2, 3, 4 made with the chisel are united where their
extremities approach each other, by means of smaller chisels, and the
connection of the preparation with the neighbouring parts is loosened
as much as possible by cutting them sideways with the chisel. Not
until then can the osseous meatus be divided in the direction from
the cranial cavity, by weaker blows of the mallet, which will drive the
chisel gradually deeper. The direction of breach 5 must be such that
its ends meet with those of 3 and 4.

The preparation, being separated from its connections with the bone,
is loosened and slightly raised, by introducing the elevator in the
breaches 1 and 2, after which any still adhering soft parts are cut
through ; the maxillary connection is divided, and the preparation
removed. This contains the inner section of the osseous meatus, the
membrana tympani, the tympanic cavity with part of the mastoid
cells, and the whole of the labyrinth as well as the auditory and
facial nerves. This method is, therefore, adapted not only for such
cases where, during life, a labyrinthine affection was diagnosed, but
also, and especially, for those so frequently occurring middle-ear
processes running their course without perforation of the membrana
tympani, in which pathological changes, mostly in the form of
adhesions and fusions, set up exclusively in the tympanic cavity,
especially in the vicinity of the ossicula, but in which Eustachian tube,
mastoid process and external meatus are not affected by the pathological process in the tympanic cavity.

The method here described may be materially simplified in cases
of undoubted affection of the labyrinth or of anchylosis of the stapes
in the fenestra ovalis ; when it will, of course, be merely a question

of removing the pyramid. For this purpose, after removal of the
tegmen tymp. with a few blows of the chisel (see chapter on Methods
of Preparation), the tympanic cavity is inspected, and having ascertained that the memb. tymp. and the malleo-incudal articulation are
normal, the tendon of the tensor tympani and the incudo-stapedial
connection are cut through. With three or four blows with the chisel
through the connection of the petrous bone with the clivus, then
through the apex, and, in the vicinity of the sinus transversus, through
the base of the pyramid, these parts may be broken off in a few
minutes and removed from the skull, the tympanic membrane with
the malleus and incus remaining behind.

2. A new method for obtaining preparations of the temporal bone
without disfiguring the body, we owe to Dr. Arthur Tschudi, whose
description we here give.

In order to take out the organs of hearing from bodies where it is
not advisable to open the cranial cavity by the removal of the skull-
cap, as usually practised, and with the object of avoiding all visible
mutilation, the following process may be employed with perfectly
satisfactory results.

Over the planum mastoideum an incision, similar to Wilde's,
is made on either side through the scalp, separating it to the
length of 10-12 cm. The cut will, therefore, as a rule, fall within
the region of the hair. The soft parts are then dissected away by
drawing the edges of the incision in the scalp forwards and backwards,
in the latter direction so far that the entire mastoid region becomes
exposed; forwards, by separation of the cartilaginous portion of the
meatus, and consequently detachment of the auricle up to about the
middle of the zygomatic arch.

By means of an American drill, with a screw-head, an opening is
made in the squamous portion of the temporal bone on one side at
a height of 4 cm. above the external meatus, which, corresponding
to the size of the drill, measures 9 mm. in diameter. The instrument,
which is double-edged, has a length of 23 cm., and may thus be driven
transversely through the entire skull to penetrate the squamous
portion of the temporal bone on the opposite side. The drill can now
be taken out of the screw-head, and the blade of a saw is then fixed
in at the posterior extremity of the metal borer, where there is a slit,
through which a peg passes. The saw-blade, made out of a watch-
spring, is 5 mm. wide and 32-33 cm. long. At its ends it is filed
out, so as to form hooks, which admit of its being fixed in the above-
mentioned pegs in the slits of the borer. The blade of the saw thus
fastened is then pulled through the bore-hole by means of the drill,
in the same way as drainage tubes are drawn through fistulous

passages. Next the saw-blade is set. The frame of the saw, specially constructed for this purpose, is rectangular, has a span of 37 cm., measures in breadth 14 mm., in thickness 4 mm., and in depth 11 cm.; it is provided with screw fastenings for tightening it in the handles, both of which are used in sawing. The work of sawing is done starting at the bore-holes, proceeding in a curve forwards and backwards round the entire exposed surface described above. While doing this the two persons sawing must guide the saw with uniform strokes, first towards the front in a gentle curve in the direction of the zygomatic arch, which is divided on the passage of the saw backwards; then directing it towards the ascending ramus of the inferior maxilla, this also is divided. The cut leaves the insertion of the tendon of the temporal muscle intact, thus securing a support to the inferior maxilla, which is a matter of consideration in regard to any possible disfigurement. When the lower jaw is cut through, the blade of the saw is loosened from the frame and withdrawn; as before, it is then pulled, by means of the metal borer, through the bore-hole, to be again set in the frame, so as to enable the operators to commence the cut in a backward direction.

This simply passes in a curve round the mastoid process, and below it joins the first cut. The bone and soft parts being cut all round, the cylinder can now be easily removed. It must be borne in mind, however, that it will be much easier to do so if the surface round which the cut has been carried be larger on one side than on the opposite one, so as to obtain, not exactly a cylindrical, but a slightly conical preparation, which is to be driven out base forward. The mere pressure of the finger is often sufficient to accomplish this. To avoid the somewhat tedious circular cut, the preparation may also be obtained in the form of a wedge. In this case two bore-holes must be made on each side, one 4 cm. perpendicularly above the maxillary joint, the other at the same height, a little behind the mastoid process. The saw is now introduced, and the two bore-holes are united by a horizontal cut, and converging cuts carried from them in a backward direction (in a curve round the mastoid process) till they meet.

The cutaneous incision must, in such a case, be made T-shape, unless the scalp, as is usual in post-mortem examinations, can be separated from one ear to the other and turned back.

The preparation thus obtained contains both petrous bones in connection. Anteriorly the cut passes through the sella turcica and divides the Eustachian tubes in such a way that about half a centimetre of the cartilaginous portion of the tube, as well as the entire maxillary articulation, remain in the preparation, while the condyle is separated.

2

Posteriorly the cut meets the base of the skull at the posterior periphery of the occipital foramen 1 cm. behind the articular surface for the atlas. Thus there are in the preparation the two pyramids, together with the mastoid process, the terminal portion of the sinus sigmoideus, as well as the whole of the clivus, a considerable part of the cartilaginous Eustachian tube, and finally the major portion of the cartilaginous external meatus.

The chief advantage of this mode of preparation consists in this : that after the removal of the petrous bones the roof of the skull is kept intact and the outward shape of the skull preserved, so that after filling up the temporal region with paper or such-like, to prevent a sinking in of the ears, the incisions in the scalp being also stitched up, any disfigurement of the body is avoided.

It is also to be observed that it is difficult, when boring through the temporal bones, to reach the precise spot on the opposite side; it is, therefore, advisable to drill a hole in each temporal bone separately, marking by exact measurement the spot from the external meatus and from the zygomatic arch. When the boring has been completed it is easy, by introducing the borer into one of the openings, to find the other.

To remove the organs of hearing from the cranial cavity of the new-born infant, neither chisel nor saw is required. As the sutures between the various bones of the skull are not yet ossified, and the bones themselves still spongy and offer little resistance, a stout pair of scissors is sufficient to cut out those parts of the organ from the base of the skull which are intended for the preparation. In the case of children who have reached the age of several weeks, a stout pair of scissors alone will not accomplish this, and the removal of the organs can only be effected by the use of strong bone scissors, forceps, and sometimes the chisel.

3. EXAMINATION OF THE CRANIAL CAVITY AND BRAIN IN MIDDLE EAR SUPPURATIONS, AND CEREBRAL DISTURBANCES OF HEARING TERMINATING FATALLY.

In those morbid processes in the organ of hearing where death is the result of extension of a suppuration of the middle ear to the cranial cavity, or where, during life, a cerebral derangement of hearing was diagnosed, it is necessary, after opening the cranial cavity, to take note of the changes occurring in the meninges, in the cerebral sinuses, and in the brain itself.

Should, during life, symptoms of meningeal or cerebral affection have existed, it is necessary, after removal of the calvaria, to test the degree of tension of the dura mater.   In diffuse meningitis of the convexity of the brain, at times also in basilar meningitis, especially when combined with purulent or serous effusion in the ventricles of the brain, further, in acute and chronic hydrocephalus, the dura mater over the cerebral hemispheres is found tightly stretched and little yielding.   On the other hand, in abscesses of the cerebrum, particularly when extensive and situated superficially, a frequently not inconsiderable degree of relaxation and pliancy is found in the corresponding parts of the dura mater.

When the latter has been removed in the usual manner from the upper surface of the cerebral hemispheres, note should be taken of the degree and extent of the hyperæmia and serous moistness, or of the purulent infiltration of the pia mater and subarachnoid spaces, and of any casual fluctuation which may be perceived on lightly touching the surface of the brain, and which frequently occurs in superficial cerebral abscesses, and in hydrocephalus.   Next, the anterior lobes of the cerebral hemispheres are raised up with the left hand, and the olfactory and optic nerves, both carotid arteries, the infundibulum, and the oculo-motor nerves are cut through in succession.   Should, upon raising the anterior lobe, any purulent infiltration be found on the under surface of the brain, the middle lobe must be raised with great care, in order to note any existing adhesions between the dura mater and the brain in the middle fossa of the skull, especially in the region of the tegmen tymp.

Now follows, on either side, the division in the usual manner of the tentorium cerebelli from the proc. clin. postic. along the upper edge of the pyramid, the trochlear and trigeminus nerves being cut at the same time.   Gradually raising the cerebellum and medulla oblongata from before backwards, we are enabled to cut in succession, as they come into view, first, the abducent, facial, and auditory nerves, further backwards the glosso-pharyngeal, the vagi, accessory nerves of Willis, and the hypoglossal.   It now suffices to separate the connection of the medulla oblongata with the spinal cord, with a scalpel carried along the clivus through the occipital foramen into the vertebral canal, in order to allow the entire brain to be removed from the cranial cavity.

It should, however, be remarked that in cases where, during life, symptoms of a cerebellar abscess have existed, attention must be directed, when lifting out the cerebellum, whether, and at what point, an adhesion of the latter with the posterior wall of the pyramid exists.

Such adhesions between the dura mater and brain in the middle and posterior cranial fossæ are not infrequently found, when a middle ear suppuration breaks through at the tegmen tymp., at the antrum mastoid., and at the posterior pyramidal wall. They are the result of inflammation which precedes the breaking through of the suppuration on the inner surface of the dura mater, and which from here spreads to the adjoining arachnoid membrane, pia mater, and the brain substance. In the great majority of cases these adhesions are met with in otitic brain abscesses, less frequently in diffuse meningitis, unaccompanied by any such abscess, and affect mostly the immediate neighbourhood of the breach in the bone, and of the opening of the abscess frequently corresponding with it. After removing the brain from the cranial cavity, the parts at its lower surface, corresponding to the breach in the bone, are next minutely examined, and the intensity and extent of the purulent infiltration at the base of the brain determined. If the brain, at the spot where the bone appears ulcerated, should show a discoloured depressed place, where, on slightly pressing the neighbouring parts, pus oozes out, the corresponding lobe should be cut through longitudinally, so as to ascertain the size and seat of the abscess, its contents, the condition of its walls, and of the neighbouring brain substance. Not infrequently the abscess communicates directly with the seat of the suppuration in the temporal bone; sometimes, however, even on a most minute examination, no connection between these two purulent foci can be found to exist. Finally the ventricles of the brain must be closely examined. They very often contain, in consecutive middle ear suppurations which have extended to the cranial cavity, cloudy purulent fluid, their walls appear injected, ecchymosed, and in places softened.

After noting down the result of the examination of the brain, the base of the skull should once more be closely examined before the organs of hearing are removed, in order to ascertain whether there are any changes in the dura mater within the region of the temporal bone, and to what extent it may be ulcerated, before stripping it off. The region above the tegm. tymp. and antrum mastoid., sometimes also the superior and posterior walls of the pyramid, appear dark-blue, livid, discoloured, loosened, and infiltrated with pus; or white and yellowish, if caseous or cholesteatomatous masses in the temporal bone are lying upon the external surface of the dura mater. Where the latter has been broken through in the direction from the temporal bone, there is mostly found a gap bounded by jagged edges, which is filled with pus or caseous masses; less frequently multiple roundish perforations are met with in the dura mater, which is thickened by infiltration. In order to ascertain the extent of the defect in the osseous roof of the tympanic cavity, the dura mater should be detached in the direction from the squamous portion of the temporal bone as far as the posterior superior edge of the pyramid (sin. petrosus superior), without, however, entirely separating it from the preparation. In the same manner, if there be a breaking through at the posterior surface of the pyramid, the dura mater may be detached from the inferior towards the superior edge of the pyramid. Only in this manner can it be shown whether the gap in the bone corresponds with that in the dura mater, or whether it is somewhat further removed from it. In the latter case the two openings communicate, by means of a channel running in a slanting direction through the thickened dura mater. By holding the latter up to the light, it is possible to judge, from the degree of opacity in the vicinity of the place of rupture, to what extent it has been infiltrated and thickened. In making post-mortems on children who have died of meningitis, or epidemic cerebro-spinal meningitis, the processes of the dura mater, which penetrate into the sutures and fissures of the temporal bone, especially of those vascular cones of connective tissue which enter the hiatus subarcuatus, must be more minutely examined, since the purulent inflammation sometimes finds its way through these from the spaces in the middle ear to the cranial cavity (v. Tröltsch, Lucæ).

In a case of sinus phlebitis terminating fatally, the sin. longitudinalis superior should first of all be split, to ascertain the condition of the coagula found in it. Then, after removing the brain from the cranial cavity, the sin. transversus is opened with scissors and pincette, from the foramen jugulare to the confluens sinuum, and the divided sinus wall turned over to either side to obtain a clear view into the sinus. The condition of the clots found here should first be examined, then the state of the sinus wall, whether it be smooth, or inflamed and covered with clots adhering to it; finally, whether the wall of the sinus adjacent to the bone has been broken through at any spot. With every thrombosis in the sin. transversus, it is necessary to open at the same time the sin. petrosus superior running along the posterior superior edge of the pyramid; likewise, the sin. petrosus inferior, which passes along the posterior inferior edge; and if pathological thrombi are found there, the examination should be extended also to the sin. cavernosus. The latter should always be done when, during life, symptoms of phlebitis of the sin. cavernosus, inflammation of the retrobulbar connective tissue, exophthalmus œdema, or erysipelas of the eyelids had existed. Should the thrombous formation extend downwards beyond the foramen jugulare, then the examination of the vena jugularis should not be neglected, so as to ascertain the extent of the inflammation and coagulation towards the lower parts of the vein, especially if, during life, a painful cord-like formation in the course of the jugular vein was proved to exist.

When the venous sinus wall has been detached from the underlying bone, it is possible to judge of the position, size, and form of the existing ulceration. This is found, in the majority of cases, in that portion of the sinus transversus which borders on the mastoid cells, and from here a probe may be passed either into the cavity of the mastoid process, or into the tympanic cavity direct. Oftener, however, even in severe sinus phlebitis, no trace of any ulceration in the bone is found, but, corresponding to the inflamed wall of the vein, a livid or yellowish-green discolouration of the bone, showing the symptoms of a septic inflammation, which has spread from the tympanic cavity or mastoid cells to the sinus. Sometimes it is by means of the fine venous trunks discharging from the mastoid cells into the sinus that this septic inflammation reaches the sinus wall.

In deafness the result of hydrocephalus, it is necessary to examine the floor of the sinus rhomboideus and the condition of the striæ acusticæ, which not infrequently are found completely obliterated; further the auditory nerve, which appears atrophied, flattened, or, as I have observed in a case of chronic hydrocephalus of high degree, combined with atrophy of the brain, lying, in the form of a slender cord, within the internal meatus, which was dilated to several times its natural size.

In disturbances of hearing consequent on tumours of the brain, attention should be directed to the relations of the tumour to the auditory nerve and the medulla oblongata. To gain information upon this we must, previous to cutting through the cerebral nerves which pass from the medulla, slightly raise the latter and the cerebellum, cutting, if necessary, the hemispheres partially away, by which the displacement of the brain, of the medulla oblongata, and of the nerves caused by the tumour, may be plainly seen.

It depends upon the size of the tumour, its close connection with the brain or with the petrous bone, whether it should be removed with the brain, or left in connection with the petrous bone. The latter is always done when the new formation invades the internal meatus, and when the auditory nerve passes through the tumour and appears completely imbedded in it.

In dissections on deaf-mutes it is also necessary to subject the brain to a close examination, and special note should be taken of any changes occurring in the ventricles, in the sinus rhomboideus, and in the auditory nerve trunks. The medulla oblongata should, for the purpose of future microscopic examination of the nuclei and roots of the acoustic nerve, be placed in a weak solution of chromic acid, to be hardened. Considering the relations which have lately been proved to exist between the acusticus and the temporal lobe, it might be advisable, in the dissection on deaf-mutes, to remove that part of the brain, for the purpose of microscopic examination.

In conclusion, let it be observed that also in fractures of the skull extending to the temporal bone, before removing the organs of hearing, a close examination of the dura mater and the cranial bones should be made, in order to determine the extent and direction of the fissures in the skull, and the rents in the dura mater.

III.

DISSECTION OF THE MACERATED TEMPORAL BONE.

AN exact knowledge of the temporal bone, which forms the osseous framework of the auditory apparatus, is indispensable for the study of the anatomy of the ear. Before beginning the work of preparation, the anatomical details of the temporal bone must first of all be studied not only on its external surface, but also on good sections. As there are material differences of form between the temporal bone of the new-born infant and that of the adult, a comparative study of both is of great advantage, since thereby we obtain information, not only as to the growth of the bone, but also as to certain pathological occurrences in the ear during the first years of life.

1. DISSECTION OF THE TEMPORAL BONE IN THE NEW-BORN INFANT.

No bone of the skeleton undergoes after birth such great changes of form as the temporal bone. Its growth can, therefore, only be studied on a series of preparations, commencing with the temporal bone of the new-born infant, and followed, in succession, by those of children aged respectively three months, six months, nine months, twelve months, up to five years.

The macerated temporal bone of the new-born infant (Figs. 20 and 21) can be divided without difficulty into three pieces; pars squamosa, pars tympanica, and pars petrosa, which, having in fœtal life a separate development, are in the new-born infant but loosely connected, and do not become intimately united until the first few years of life.

The details of the temporal bone of the new-born infant can only be studied from its various parts after anatomical dissection.

This is not attended by any special difficulty; still, in order to avoid splinterings, certain precautions ought to be observed.

We begin by detaching the tympanic ring (annulus tymp., a), inserting a small knife rounded off at the point, or the blade of an ordinary penknife, into the fissura petro-tympanica (fissura Glaseri), which lies between the anterior crus of the ring and the tegmen tymp., so as to loosen, by slight lateral pressure, the anterior crus, which has its attachment immediately behind the process. articularis post., from the pars squamosa. If, now, the connection between the lower and posterior periphery of the ring and the corresponding edge of the pars petrosa et mastoidea be loosened, by pushing the point of the knife into the loosely constructed suture, a gentle pressure sideways will free the upper end of the posterior crus of the ring from the squamous portion, whereby the separation, intact, of the annulus is effected.

This is followed by separating the pars squamosa from the pars petrosa, which are united on the inner side of the temporal bone (Fig. 21) by means of the sutura petro-squamosa (su su'), and on the outer surface through the sutura mastoideo-squamosa (Fig. 20, d d).

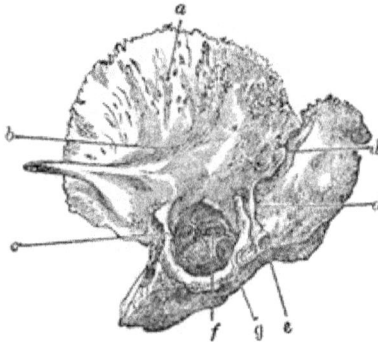

FIG. 20. Outer view of the temporal bone of a new-born infant. (Left ear.)—a = upper portion of the squama; b = its inferior portion below the linea temporalis passing backwards; c = annulus tympanicus; d d = suture between the squamous portion and the mastoid process, extending to the foramen stylo-mastoideum; e = foramen stylo-mastoideum'; f = fenestra ovalis; g = fenestra rotunda.

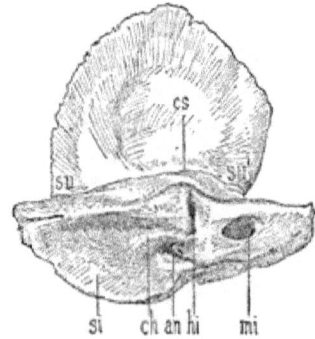

FIG. 21.—View of the inner side of the temporal bone of a new-born infant. (Left ear.)—su su' = sutura petro-squamosa; cs = tegmen tymp.; mi = meat. audit. int.; hi = fossa subarcuata; an = fissure of the aquæduct. vestib.; ch = elevation of the horizontal semicircular canal; si = place of the later sulcus transversus.

By cautiously pushing the blade of the knife into the latter, and by gentle lateral movements of the instrument, the connection of the two bones can be loosened, so that by moderate pressure on the inner side of the squamous portion the connection at the sutura petro-squamosa may also be loosened, and the pars squamosa isolated.

The temporal bone, thus divided into its three constituent parts, can now be accurately examined in all directions; and without entering into an exhaustive treatment of the subject, we will give a short description of the different pieces of the os temporale in the new-born infant as far as this appears necessary for understanding the anatomical details of the temporal bone in the adult, and for technical work.

The tympanic ring (annulus tymp., Figs. 22 and 23) presents an osseous frame, open above and in front for the reception of the membrana tympani, and attached by its ends to the anterior lower part of the external surface of the squamous portion, while its lower and posterior circumference unites with the lower external edge of the pars petrosa, and with the mastoid portion of the latter. On the outer side of the ring are two projecting tubercles, tuberculum tymp. antic. (a) and postic. (p) (Zuckerkandl) which should be noticed, because the development of the osseous meatus out of the tympanic ring commences at these two spots. The inner side of the ring (Fig. 23) presents a groove (sulcus tympanicus, st), intended for the reception of the tympanic membrane. The inner lip of the groove at the anterior crus of the ring runs above into a point (s) directed backwards and inwards, which Henle calls spina tymp. post., in contradistinction to the spina tymp. ant. (t), which, lying somewhat lower, projects at the anterior edge of the crus. The two processes are joined by an

Fig. 22.—Outer side of the annulus tympanicus. (Left ear.)—a = tuberc. tymp. anter. ; p = tuberc. tymp. post. (Zuckerkandl.)

Fig. 23.—Inner side of the annulus tympanicus. (Left ear.) — st = sulcus tymp. ; p = tuberc. tymp. post. ; t = spina tymp. ant. ; s = spina tymp. post. (Henle). Between the two, the crista spinarum, with the sulcus malleolaris passing below.

inwardly-projecting ridge (crista spinarum, st), below which the sulcus malleolaris, serving for the reception of the proc. long. mallei, passes in an oblique direction forwards and downwards. The fissure situated between this portion of the anterior crus and the anterior outer edge of the tegm. tymp. is called fissura petro-tympanica (fissura Glaseri), through which pass part of the ligam. mallei ant., the art. tymp. inf., and the chorda tymp.

The second piece of the temporal bone in the new-born infant, the pars squamosa (Figs. 24 and 25), shows on the anterior part of its outer surface the slender processus zygomaticus, the root of which is bounded below by the still faintly marked and shallow articular fossa of the inferior maxilla. The upper edge of the proc. zygomaticus is extended as a slightly curved ridge (linea temporalis), which divides the squamous portion into an upper and a lower segment. The latter portion, which has the form of a triangle with its apex directed

downwards, lies below the linea temporalis, and claims our special
attention, since it takes an essential part in the later development
of the osseous meatus and of the mastoid process.   The anterior edge
and section of this triangle are divided into three parts.   The uppermost part (a r), lying behind the proc. articul., serves for the apposition
of the squamiform end of the anterior crus of the tympanic ring; the
middle part (r r') forms, in common with the annulus, the so-called
incisura Rivini (margo tymp. of the squamous portion); while the
posterior third (r' l), with which the extremity of the posterior crus
of the ring coalesces, is joined to the tuberc. mast. of the pars
mastoidea.   The posterior section of the triangular part of the squama
(l q), which, through the sutura mast. squamosa (Fig. 20, d d), is
united with the mastoid portion of the pyramid, forms the outer
covering of the antrum mast.; and as especially the lower portion,

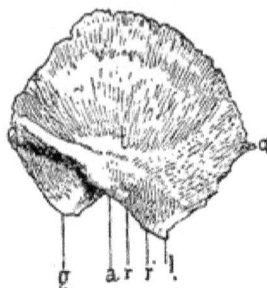

FIG. 24.—Outer surface of the pars squamosa
of a new-born infant. (Left ear.)—g =
anterior lower edge of the squama; above
it the shallow cavit. glenoid.; a r = place
of attachment of the anterior crus of the
ring to the squama; r r' = margo tymp.
squamæ; r' l = lower point of the pars
squamosa in apposition with the tuberc.
mast.; l q = posterior edge of the pars
squamosa, forming the sutura mast. squamosa.

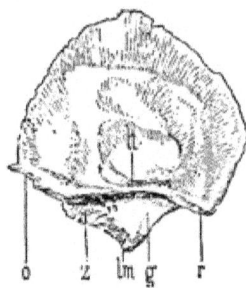

FIG. 25.—Inner view of the pars squamosa
of a new-born infant. (Left ear.)—lt =
inner projecting lamella of the squama;
lm = inferior triangle of the squamous
portion; g = smooth part of this lamella;
z = its cellular portion; o = posterior
boundary of the sutura petro-squamosa;
r = anterior, groove-like border of the
lamella.

adjoining the tuberc. mast., takes part, during the growth of the
mastoid process, in the formation of the anterior section of the latter,
Zuckerkandl has designated it lamina mast. externa.

    On the inner side of the squamous portion (Fig. 25) the lower part
of the surface is formed into a triangle by a projecting horizontal
jagged ridge (o r).   The upper lamella of this triangle (o r) is united,
by means of the sutura petro-squamosa, with the tegmen tymp. et
mast. (Fig. 26), arising from the upper surface* of the pyramid.   The
posterior and middle portion of this edge is sharpened as far as the
anterior superior angle of the triangle, and overlapped by the corre-

    * Syn., inner anterior surface (Henle), anterior superior surface (Luschka).

sponding sharpened plate of the tegmen. tymp. (Fig. 26, l). The
anterior broader part (Fig. 25, r), on the other hand, appears
depressed into a groove to receive the correspondingly thickened
edge of the tegmen tymp.

The lower lip of this groove (proc. inf. tegm. tymp., Kirchner)
is wedged in between the anterior lower border of the squama and
the anterior crus of the ring, and is visible on the temporal bones of
the new-born infant and adult as a long ridge between the squama
and the upper edge of the os tymp. (Fig. 29, g). The fissure
remaining between this lip and the os tymp. is the fissura petro-
tymp., or fissura Glaseri. This is connected in the new-born infant
(Fig. 28) with that triangular gap which lies in the anterior angle
of the temporal bone, called by Henle canalis musculo-tubarius, and
which is divided by the ledge of bone projecting from the median
wall of this canal into two not quite distinct canals, the lower and
wider of which represents the tuba Eustach. ossea, the upper narrower

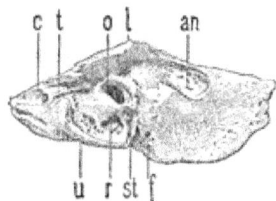

FIG. 26.—Pars petrosa of a new-born infant, after removal of the pars squamosa et tympanica.
—o = fenestra ovalis ; r = fenestra rotunda ; st = sinus tymp. ; u = outer border of
the lower surface of the pyramid ; f = foramen stylo-mastoid. ; l = lamella of the
tegmen tymp. ; an = antrum mastoid. ; t = ost. tymp. tubæ Eust. ; c = canalis caroticus.

one the semicanalis pro tensore. The anterior corner of the tegm.
tymp. sometimes shows, at the spot where the upper and lower lips
of its lateral edge meet, a roundish surface about the size of 1 mm.
covered with cartilage and surrounded by a low elevation.

The triangular surface of the squama situated below the jagged
ledge is smooth at its anterior portion, but cellular at the posterior.
The anterior smooth portion (g) forms, with the jagged ledge overlapping it, a niche in the upper tympanic space, in which the malleus
and incus are lodged, while the posterior cellular portion (z) of the
triangle represents, as already mentioned, the outer roof-plate of the
antrum mast.

The third piece of the temporal bone of the new-born infant, the
pars petrosa (Fig. 26), contains in its anterior and middle part the
cavity of the labyrinth; in the posterior, the first indications of the
mastoid process, and the antrum mast. On its external surface there
are to be noticed the fenestra ovalis (o) and rotunda (r) ; the

promontory projecting in front of these; the portion of the facial
canal running obliquely backwards and downwards above the fenestra
ovalis, with the prominence of the horizontal semicircular canal
adjoining it posteriorly; the eminentia stapedii projecting behind
the fen. ovalis; and, in the anterior part of this wall, the lower inner
limit of the osseous Eustachian tube, with the semicanalis pro tensore
tymp. (t) above it. The longish oval antrum mast. (an), situated
behind the tympanic cavity, measures in its long diameter 9-11 mm.,
and in height 7-9 mm. The mostly smooth but often also porous
inner and lower wall of the antrum does not attain until later, during
the growth of the mastoid process, a distinctly cellular structure.

The inner wall of the tympanic cavity and the antrum mast.
are roofed over by the previously-mentioned triangular lamella (l)
(Fig. 26), which proceeds from the anterior superior wall of the
pyramid, and which with its sharp edge pushes the tegmen. tymp.
et mast. over the inner lamella of the pars squamosa (Fig. 25, lt),
projecting towards it from the opposite side. The traces of this
superposition of the two lamellæ can be shown to exist on frontal
sections of temporal bones in adults, and on the sections of the
superior wall of the osseous meatus and of the antrum. A groove
(sulcus N. petr. sup. maj.) commencing above the anterior orifice of
the canal. tensor.-tymp., or at the canalis caroticus, and running
backwards and laterally along the boundary of the anterior edge of
the pyramid and of the tegmen tymp., leads to the hiat. canal.
Fallopii. The facial canal, at this spot uncovered, appears here, on
an average, larger in the new-born infant than in the adult.

Noteworthy among the details on the posterior surface of the
pyramid (Fig. 21) are, besides the porus acusticus intern. (mi) in the
anterior segment, the niche (fossa subarcuata—hi) which, lying below
the superior semicircular canal, interrupts the upper margin of the
pyramid, and is occupied by a vascular process of connective tissue
proceeding from the dura mater (Wagenhäuser), but during growth
becomes so obliterated that in the temporal bone of the adult only
traces of it are to be found; secondly, the superior crus of the
posterior semicircular canal (ch), which projects at the same height
with the porous acusticus intern. behind the inner crus of the canal.
semicirc. super., as a roundish eminence; and, lastly, the depressed
orifice of the aqæd. vestibuli, situated below this elevation. The
slightly concave surface of the most posterior section of this wall of
the pyramid (Fig. 21, si), shows as yet scarcely any indication of the
future sulc. sinus transversus.

For the study of the topography of the temporal bone in the new-born infant a few sections through the entire bone are required. Three in particular are absolutely necessary for understanding the respective details in the temporal bone of the adult. One of these is a frontal cut passing vertically through the posterior part of the squama, through the tuberculum mast. and the foramen stylo-mast., and dividing, in addition to the squamous and mastoid portions, the posterior part of the petrous bone. It shows (Fig. 27) the outlines of the antr. mast., its height and transverse diameter, and the share the pars squamosa et petrosa take in the formation of its walls. Very plainly visible in such sections is the overlapping of the lamella (t), proceeding from the superior surface of the pyramid, and of the lamin. mast. int. of the squama (l); also the apposition of the cellular portion of the squama (a) to the tuberc. mast. (tu). The second frontal cut (Figs. 31, 32) passes through a plane, which lies on the

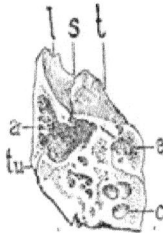

FIG. 27.—Frontal section through the antrum mastoid. of a new-born infant: posterior view. (Right ear.)—a = outer lamella of the antrum belonging to the squamous portion ; l = lamin. mast. int. of the pars squamos. sutura petro-squamosa ; t = tegm. tymp. ; tu = tuberculum mastoid. ; e = fossa subarcuata ; c = cochlea.

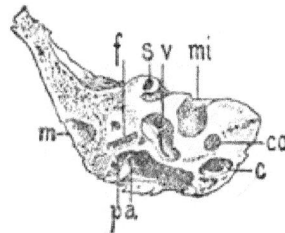

FIG. 28.—Horizontal section through the temporal bone of a new-born infant. (Right ear.)—a = section of the posterior crus of the ring ; p = section of the sheath of the proc. styloid. ; m = lower end of antr. mastoid. ; f = section of the canal. facialis ; v = vestibule ; s = section of the canal. semic. sup. and of the fossa subarcuat. ; co = cochlea ; mi = meat. intern. ; c = canal. caroticus.

one side through the greatest diameter of height of the squama, the Rivinian segment, the posterior crus of the annulus (midway between the entrance of the canal. carotic. and the foram. stylo-mast.), and on the other side through the internal auditory meatus. It meets the promontorial wall close in front of the niche of the fenestra ovalis. At the anterior part of this section (Fig. 32) we notice the transition of the tympanic cavity into the canalis musculo-tubarius ; at the posterior piece (Fig. 32), the passage of the upper tympanic space (c) into the antr. mast. This section, an illustration of which is given in the following chapter, is, as we shall see, also of importance for the study of the growth of the external meatus. The horizontal cut through the temporal bone (Fig. 28) falls in a plane lying through the tubercul. mastoid., the lower wall of the osseous Eustachian tube, and the

internal meatus.   It effectually shows the mode of apposition of the
annulus tymp. to the edge of the lower and posterior walls of the
tympanic cavity (a) ; the relative position of the canal. carot. (c) to
the osseous Eustachian tube, and to the anterior wall of the tympanic
cavity; the shape of the pars mastoidea in transverse section ; at the
posterior boundary of the cavum tymp., the position of the facial
canal (f) and of the sheath of the styloid process (p); and lastly, at
the inner edge of the pars petrosa, the section of the hiat. subarcuatus (s).

The sections through the petrous bone of the new-born infant,
which are to represent the cavities and canals of the osseous labyrinth

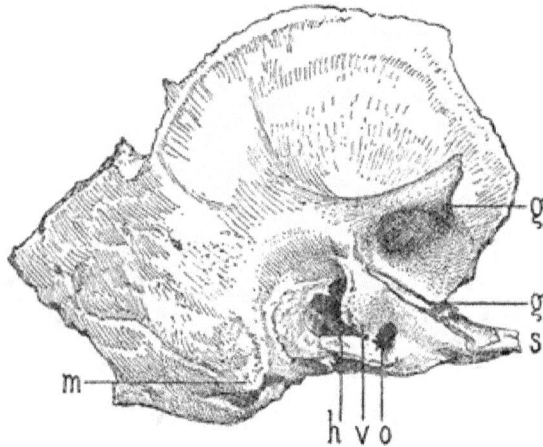

Fig. 29.—Temporal bone of a child one year old.   (Right ear.)—v = osseous growth arising
from the tubercul. ant. of the tympanic ring ; h = osseous growth from the tubercul.
postic. ; o = gap left between the anterior and posterior osseous growths, meeting at the
anterior lower wall of the auditory canal ; g g = cavitas glenoidal. with the root of the
proc. zygomat. : the lower g marks likewise the inferior lip of the tegm. tymp. thrust
between the squama and the os tymp. ; s = apex of the pyramid ; m = proc. mastoid.

for the purpose of study, will be discussed more fully in the chapter
' Preparation of the Osseous Labyrinth.'

Those who desire to gain information as to the various stages of
the growth of the temporal bone as they present themselves in early
infancy up to the age of the adult, may with advantage make use of a
collection of skulls arranged according to age, such as are to be found
in great anatomical institutions.   A model collection of this kind is
that of the Anatomical Museum of Vienna, the inspection of which
will be found very instructive.

From the examination of a number of skulls arranged according
to age, we generally find that the growth of that part of the osseous
meatus which proceeds from the pars tymp. progresses most rapidly

and terminates soonest; that, however, the upper wall of the meatus, formed by the pars squamosa (horizontal portion) does not attain until later its permanent length and width; and that the mastoid process is fully developed last—often not until puberty, and is even after that time subject to internal changes.

Should a series of skulls or temporal bones arranged according to age not be accessible, the development of the osseous meatus after birth may be followed on three or four temporal bones: of the new-born infant (Fig. 20), of children aged respectively one and two years (Fig. 29), and of the adult (Fig. 30).

The growth of the osseous meatus, in the formation of which the squamous and tympanic portions of the temporal bone chiefly partici-pate, proceeds in the following manner: As early as a few weeks after birth an increase of substance may be observed on the tubercles at the lateral side of the tympanic ring, which have been called tuberc. antic.

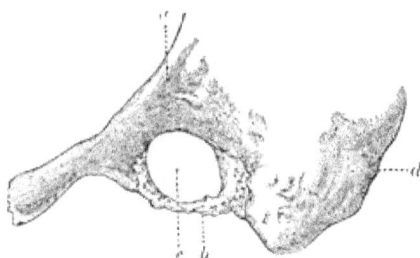

Fig. 30. — Outer view of the osseous meatus of an adult. (Left ear.)—a = horizontal portion of the squama, upper wall of the meatus; b = pars tympanica; c = lumen of the auditory canal; d = mastoid process.

and postic. (Fig. 22, a p, Zuckerkandl). The rapid growth of these tubercles (Fig. 29, v h), and the simultaneous increase of substance in the whole of the tympanic ring, mostly lead as early as the first year of life to a bridge-like union between them. There remains between the lower periphery of the annulus and the osseous bridge, which forms the outer section of the anterior and lower walls of the meatus a gap (o), varying in size, and closed by fibrous tissue, and which frequently up to the third year of life is filled with bone; often, how-ever, it persists up to the fifth or sixth year, and even sometimes throughout life (Ossificationslücken of Arnold and v. Tröltsch).

On the temporal bone of the adult the part of the osseous meatus formed by the pars tymp. presents itself as a groove-like convoluted osseous plate (Fig. 30, b), which looks as though pushed from below into the sulcus, facing downwards, and which is formed by the hori-zontal portion of the squama (a), the process. articul. post., and the mastoid process.* The limits of apposition of the os tymp. to the squamous and mastoid portions are generally recognisable by the trace of a suture, rarely quite obliterated.

The formation of the superior wall of the meatus out of the squamous portion proceeds in such a manner that the squama proper above the linea temporalis (Fig. 31) retains its position; while that part (p p, superficies meat. Schwalbe) which is situated below this line, between the process. artic. postic. and the tuberc. mastoid., gradually assumes during growth a horizontal position.

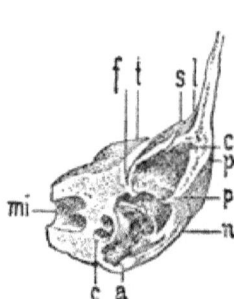

FIG. 31. — Frontal section through the middle of the temporal bone of a new-born infant: posterior view. (Left ear.)—p p = lower portion of the pars squamosa (superficies meat. ext.); n = attachment of the posterior crus of the ring to the pars mastoid. oss. petros.; a = suture between the lower periphery of the annulus and the margin of the pars petrosa; l = lamin. int. of the pars squamosa; s = sutura petro - squamosa; t = tegmen tymp.; f = hiat. can. Fallopii; c = upper tympanic space; c = cochlea; mi = meat. intern.

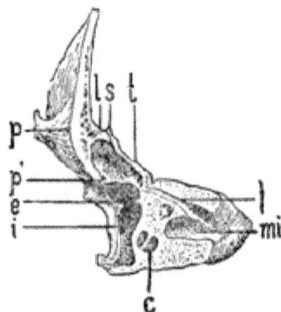

FIG. 32. — Frontal section through the middle of the temporal bone of a new-born infant: anterior view. (Left ear.)—p p' = lower portion of the pars squamosa (superficies meat. ext.); l = lamin. int. of the pars squamosa; s = sutura petro - squamosa; t = tegm. tympan.; e = fissura petro - tympan. (Glaseri); i =canalis musculo-tubarius; c = cochlea; mi = meat. int.; f = canalis facialis.

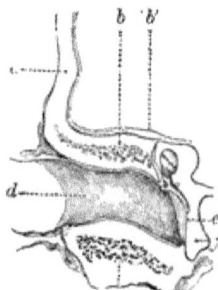

FIG. 33.—Vertical (frontal) section through the osseous meatus and the tympanic cavity. (Left ear.) — a = squama; b' = horizontal portion of the pars squamosa; b = upper wall of the osseous meatus; c = lower wall of the osseous meatus; d = meatus; e = tympanic membrane with the malleus; f = tympanic cavity.

If the sections of the temporal bone of the new-born infant (Figs. 31, 32) and those of the adult (Fig. 33), as shown in the accompanying illustrations, be placed side by side, it is easy to see that the piece marked p p' of the pars squamosa (Fig. 32) is the one which becomes during growth the portion (b) shown in Fig. 33, forming the upper wall of the meatus, and standing at a right angle to the squama proper. After what has been said, the anterior and lower walls of the osseous meatus are thus formed exclusively out of

* Du Verney, Traité de l'organe de l'ouie, 1731.

the os tympan., the upper wall almost entirely by the pars squamosa; while in the formation of the posterior wall the tympanic, mastoid and squamous portions participate, the latter in a great measure by the lamina mastoidea, which forms the anterior surface of the mastoid process.

2. SECTIONS THROUGH THE TEMPORAL BONE OF THE ADULT.

For the study in detail of the temporal bone in the adult, a series of sections running parallel in a sagittal, frontal, and horizontal direction are made by means of a fine fret-saw (No. 12) through three different temporal bones. In order not to miss the direction of the cuts, it is well to mark them, as to their position and distance from each other, by drawing lines on the outer surface of the bone. As the sections of the cavity of the labyrinth will be discussed in the chapter

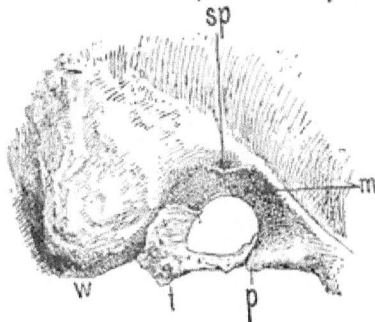

FIG. 34.—External opening of the osseous meatus on the macerated temporal bone of an adult. (Right ear.)—t = pars tymp. of the osseous meatus; m = pars squamosa (horizontal portion) of the osseous meatus; sp = spina supra meatum (Bezold); p = proc. articularis of the maxillary joint; w = mastoid process.

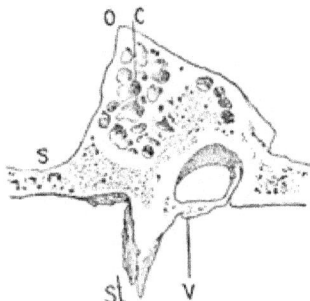

FIG. 35.—Sagittal saw-cut, 5 mm. medially from the external opening of the osseous meatus on the same preparation.—v = anterior lower wall of the osseous meatus belonging to the pars tymp.; c = pneumatic cell-spaces of the middle ear, situated along the posterior upper wall of the meatus; o = upper edge of the pyramid; s = sin. sigmoideus; st = proc. styloideus.

on the preparation of the osseous labyrinth, we will in the following chiefly consider the sections of the osseous meatus and of the middle ear.

The first series of sections through the temporal bone in a sagittal direction is made perpendicular to the axis of the external meatus. Three parallel sections, from the external orifice to the incisura Rivini, suffice to bring into view the change of the transverse section of the external meatus from without inwards, and its topographical relations to the glenoid cavity, to the mastoid cells, and to the antrum mastoideum.

As the width of the osseous meatus in the normal state is subject to very great variations, a number of temporal bones should be chosen for making the sections, which should include, besides specimens with meatuses of average width, also such having extremely wide and very narrow ones. The sections here illustrated are taken from a temporal bone with a meatus having a lumen of average width.

The first cut commences at the inner surface of the squamous portion, close to the boundary between the lower and lateral walls of the middle cranial fossa, falling thus on an average 5 mm. medially from the inferior rough edge of the pars tymp. A comparison of the transverse diameter of the external orifice of the meatus (Fig. 34) with that of the first section (Fig. 35) shows that the height of the latter has diminished by $2\frac{1}{2}$-$3\frac{1}{2}$ mm., and the transverse diameter by 2-3 mm. In the following cut, made 3 mm. medially through about the middle of the lower wall (Fig. 36), the height of the elliptic trans-

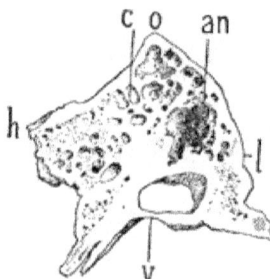

Fig. 36.—Sagittal section, 3 mm. medially from the previous one.—v = anterior lower wall of the meatus; an = lateral portion of the antrum mast.; c = pneumatic cell-spaces; o = upper edge of the pyramid; h = its posterior edge.

verse section, which is strongly inclined forward, is found but slightly less; the transverse diameter, however, shows a decrease of 2-3 mm. Lastly, on the third cut (3 mm. towards the median line) passing through the incisura Rivini and the posterior superior portion of the sulcus tymp., the height again increases by 1-$1\frac{1}{2}$ mm., while the transverse diameter (1 mm. less than on the previous section) is here narrowest.

The narrowest place (isthmus meat. ext.) of the osseous meatus, diminishing from without inwards, especially in the transverse diameter, is thus, on an average, between the posterior superior edge of the membrana tympani and the anterior inferior wall of the meatus opposite, which is here most strongly curved backwards. Sometimes the isthmus commences $1\frac{1}{2}$ mm. laterally from the posterior margin of the tympanic membrane.

The osseous meatus in the adult resembles a laterally compressed cylinder, the transverse section of which takes the form of an irregular oval with its greatest diameter strongly inclined forwards. Such being the form of the section, it is, as Bezold (*l.c.*) rightly observes, impossible to sharply define the limits of the walls of the meatus. Indeed, one may, not only on corrosion sections, but also on sagittal sections of the temporal bone, and by a glance into the meatus in a macerated skull, become convinced that, in consequence of the lateral flattening of the auditory canal, two of its walls are more strongly marked, viz., the posterior upper and the anterior lower, which meet in a short curve at the extreme points of the long diameter of the oval.

The measurements of the transverse diameter of the meatus must be made, not only on sagittal, but also on horizontal and frontal sections of the temporal bone; likewise on corrosion preparations, according to Bezold's (*l.c.*) method. When determining the dimensions from the latter, one should, notwithstanding the great advantages gained thereby, not dispense with taking the measurements of bone sections also into account. Considering the great importance of the subject, anatomists and otologists have directed their attention especially to the measurements of the auditory canal, and Von Tröltsch, Sappey, Hyrtl, and Bezold have already given such detailed information, that hardly anything new remains to be added. I can, however, on looking over these measurements, not avoid observing that I do not quite understand the mode of determining the average number from a long list of single measurements. Not infrequently, average numbers occur, which are not found in any single case of the series in question. Now, as such average measure does not exist in reality, it is, in my opinion, of no practical value. More simple does it appear to me to quote such measurements as mean results which have been found in the greatest number of preparations, and along with these to state how far, in the smaller number of preparations, these figures are exceeded or not.

The measurements made by me on sagittal sections of the meatus gave, in the majority of temporal bones, the following results: At the external opening of the osseous meatus in the vertical plane of the anterior edge of the pars tymp. a high diameter of 12 mm., transverse diameter 10 mm. In the middle of the osseous meatus, high diameter 9 mm., transverse diameter 5 mm. At the inner extremity of the superior wall (incisura Rivini), high diameter 10 mm., transverse diameter 4 mm. In the smaller number of temporal bones, with very wide and very narrow meatuses, these numbers vary between 1·3 mm. and above. Bezold found, on corrosion sections, the following average measurements: at the commencing portion of the osseous meatus, a high diameter of 8.67 mm., a transverse diameter of 6.07 mm.; at the end of the meatus in the vertical plane of the incisura Rivini, a high diameter of 8.13 mm., a transverse diameter of 4.6 mm. The reason of this perceptible difference between the results of the two measurements is probably to be found in the fact that the majority of Bezold's corrosion casts were taken from unmacerated organs of hearing, in which the lumen of the meatus, from the periosteal and cuticular lining, must appear materially less than in the macerated temporal bone.

The sagittal sections through the external meatus here illustrated (Figs. 35 and 36) give a general view of the relations between the posterior superior wall of the meatus and the pneumatic spaces of the mastoid process. As the mastoid process proper lies more laterally than the external opening of the osseous meatus, the previous cut, dividing the squama and zygomatic process, will include also the entire mastoid process (Fig. 34, w), and a part of the sinus sigmoideus.

On the first sagittal section, made medially from the external opening of the meatus (Fig. 35), appear those pneumatic cellular spaces, situated between the posterior superior wall of the meatus on the one side, and the sinus transversus, as well as the posterior surface of the pyramid, on the other side, which effect the communication of the medially placed antrum mast. with the lateral cellular spaces of the mastoid process. Their exceedingly varying number and arrangement stands, as may be seen especially on horizontal sections, in a certain relationship to the pneumatic or diploetic condition of the mastoid process.

When the series of sagittal cuts has passed the middle of the meatus (sometimes before), the lateral boundary of the antrum mastoid. is very often met with (Fig. 36). It thus extends in a lateral direction considerably beyond the boundaries of the tympanic cavity. Its position, behind and above the osseous meatus, sufficiently explains the not infrequent spreading of suppurative processes in the antrum and adjoining cellular spaces to the osseous meatus. But this anatomical relationship shows, at the same time, that the operation of opening the antrum mastoid. may also be performed from the meatus, a proceeding which was first proposed by K. Wolf,* and which, no doubt, will be more extensively practised in future than has hitherto been the case.

The sagittal cut (Fig. 37), made through the incisura Rivini and the posterior portion of the sulcus tymp., strikes the anterior wall of the meatus at a distance of 4-5 mm. from the most medially situated anterior lower portion of the sulcus tymp. It exposes that part of the antrum which lies behind the tympanic cavity and the promontorial wall of the latter. Owing to the obtuse angle which the axis

* Berl. klin. Wochenschr. 1877.

of the pyramid forms with that of the external meatus, this cut meets also the horizontal (h) and the posterior (n) semicircular canals.

This inclination forwards and inwards of the pyramid is also the cause why the sagittal sections made vertically to the axis of the meatus further inwards, strike the tympanic cavity and the pyramid in so slanting a direction that they gradually become similar to the frontal ones, which are carried perpendicularly to the axis of the pyramid.

The medial cuts of the sagittal series here described, meeting the tympanic cavity as well as the pyramid in an oblique direction, it is necessary to make a sagittal cut through the tympanic cavity on a second temporal bone, in such a way that it brings into view the pars squamosa et tymp. with the annulus tymp. on the one side, and the pars petrosa, along with the inner wall of the tympanic cavity, on the

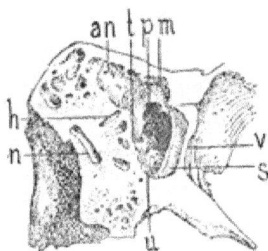

FIG. 37.—Sagittal section, 3 mm. medially from the previous one. (The section as here represented, is supposed to be turned 90° forward.)—v = section of the anterior lower wall of the meatus ; s = sulcus tymp. ; an = antrum mast. ; t p m = osseous ledge in the upper tympanic space, from the rostr. cochleare to the spina tymp. post. ; m = cavit. cap. mallei ; u = lower wall of the tympanic cavity; h n = sections of semicircular canals.

other side. The cut must thus pass in a sagittal direction through the tegmen tymp. and through the floor of the tympanic cavity, between the lower segment of the sulcus tymp. and the inner tympanic wall. As on the intact temporal bone the direction of the cut may be easily missed, it is better to make sure of it by opening the tegmen tymp. to the width of $1\frac{1}{2}$-2 mm. with a graver or forceps, in a line drawn from the anterior outlet of the canalis musculo-tubarius towards the incisura parietalis. This gives a free view into the tympanic cavity, and enables us to divide the lower tympanic wall, between the sulcus tymp. and the promontory, exactly in the middle. The temporal bone thus cut in two, shows, on the one side (Fig. 38), the outer wall of the tympanic cavity and the lateral wall of the osseous Eustachian tube (d) ; behind it, the roundish elliptic gap for the reception of the tympanic membrane, the sulc. tymp. (a) ; at the upper segment of the tympanic frame, the Rivinian notch (b), to receive Shrapnell's membrane; above that, the malleo-incudal niche (b c) and the transition of the smooth surface into the cellular wall of the antrum mastoid. The inner aspect of this section shows, when the preparation is in a normal position, the strong inclination of the tympanic frame and of the entire external wall of the tympanic cavity to the horizontal, and the acute angle which the anterior lower wall of the meatus, ascending almost perpendicularly from the lateral side of the sulc. tymp., forms with the plane of the annulus tymp.

On the inner piece of the temporal bone, sawn through in a sagittal direction (Fig. 39), we find the details of the inner wall of the tympanic cavity already mentioned on the pars petrosa of the new-born infant (p. 27). Fenestra ovalis (o) et rotunda (r), sinus tymp.,

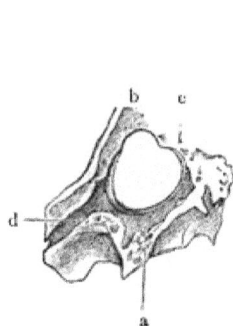

Fig. 38. — External tympanic wall with the sulcus tympanicus. (Right ear.)—a = sulcus tymp. ; b = incisura Rivini (margo tympanicus); d = lateral wall of the Eustachian tube.

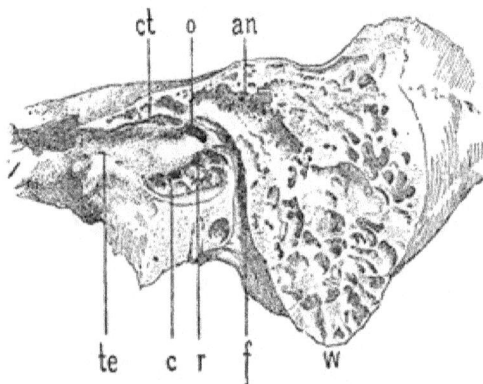

Fig. 39.—Sagittal section through the middle of the tegm. tymp. and of the lower tympanic wall. Lateral view of the inner half. (Left ear.)—te = tub. Eust. oss. ; ct = canalis pro tens. tymp. ; c = lower, ridged tympanic wall ; o = fenestra ovalis ; r = fenestra rotunda ; an = antrum mastoid. ; f = canalis Fallopii, communicating, towards the front, with the cavit. stapedii ; w = mastoid process.

promontorium, canalis pro tens. tymp. (ct), median wall of the Eustachian tube (te), eminentia staped. and the protuberance of the horizontal semicircular canal extending into the antrum mast. appear almost exactly of the same dimensions and relative positions as in the new-born infant. Merely by the increase of substance at the apex of the pyramid, at the tegmen tymp., at the lower wall of the tympanic cavity, but especially through the growth of the mastoid portion, does the pars petrosa in the adult reach, in all its dimensions, almost more than double of what it is in the new-born infant.

Now, while the details of the inner tympanic wall appertaining to the labyrinthine capsule undergo after birth scarcely any change, the canals running in the temporal bone become in the adult considerably longer, and their position is somewhat altered. Most remarkable is the increase in the dimensions of the carotid and facial canals, concurrent with the growth of the os tymp. and of the anterior point of the pyramid, these canals attaining almost double the length they have in the new-born infant.

According to my measurements, the length of the canalis caroticus in the new-born infant amounts to 10-12 mm., in the adult to 25-30 mm.; the greatest diameter at the lower opening of the canal in the new-born infant 4½-5 mm., in the adult 8-9 mm.; the greatest diameter at the upper opening of the canal in the new-born infant 3-4 mm., in the adult 6-7 mm.

The sagittal cut here described meets the middle portion of the sin. sigmoid. so far medially as to leave the entire mastoid process with the incisura mast. on the outer piece of the temporal bone. At the same time the facial canal on the posterior wall of the tympanic cavity is struck in such a way that its lowest portion also falls in the lateral piece of the divided temporal bone. If, in dividing the temporal bone, it is intended to retain the entire facial canal as far as the foramen stylo-mast. on the inner of the two pieces, the sagittal cut should pass through the lateral side of the foramen, as shown in Fig. 39, dividing thus the sulcus tymp. in such a direction as to leave a small portion of the lower wall of the meatus on the inner piece of the preparation.

Equally instructive for forming an idea as to the relative positions of the cavities and canals in the temporal bone will be found a series of sections, made in a frontal direction nearly vertically to the long axis of the pyramid, from its apex to the hindmost portion of the mastoid process. From eight to ten cuts, carried with the saw through the temporal bone at various distances, in front, 3-4 mm., in the middle and behind 5-6 mm. from each other, give a good idea of the section of the Eustachian tube and tympanic cavity, as well as of the topographical position of the labyrinthine cavity to the cavum tymp. Considering the practical importance of the middle ear, the anatomical relations of this series of sections will be briefly sketched here.

First cut (Fig. 40) about 1½ cm. behind the apex of the pyramid and about 5 mm. behind the anterior opening of the canalis musculotubarius. The osseous Eustachian tube (tu) is cut through about its middle, the lamella of the canalis pro tens. tymp. (l) in its anterior portion. The form of the section of the tube is mainly a triangle, having its base at the tegmen tymp. (tg), and its apex (tu) directed downwards. The thick lateral wall of this triangle (lateral wall of the tube) belongs to the pars tymp., the inner thin one to the canalis caroticus (c).

Second cut (Fig. 41), 3 mm. behind the first. It meets the anterior segment of the lower turn of the cochlea (co), the anterior section of the canal of the tube (tu), which is here quadrangular and generally quite separate from the canalis pro tens. tymp. (m), and bounded medially by the compact and dense wall of the labyrinth (w).

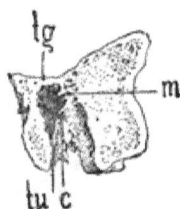

Fig. 40.—Section through the middle portion of the osseous Eustachian tube. (Right ear.)—tg = tegmen tymp.; tu = transverse section of the tub. Eust. osaea; c = its medial wall, formed by the canal. carot.; m = inferior osseous lamella of the canal. pro tens. tympani.

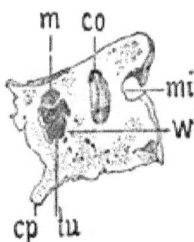

Fig. 41.—Frontal section of the temporal bone in the vicinity of the anterior segment of the lower convolution of the cochlea, 3 mm. behind the previous section. —tu = transverse section of the osseous Eustachian tube in the neighbourhood of the ostium tymp. tubæ; w = its inner compact wall; m = canal. pro tens. tymp.; co = cochlea; mi = meat. audit. int.; cp = crista tympanico-petrosa (crista petrosa).

Third cut (Fig. 42), 3 mm. behind the previous one. This meets the tympanic cavity anteriorly, between the ost. tymp. tubæ and the anterior segment of the sulcus tymp., the distance between which

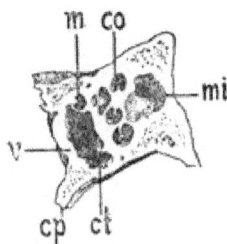

Fig. 42.—Frontal section, 3 mm. behind the previous one, through the axis of the cochlea. — ct = transverse section of the tympanic cavity, immediately behind the ost. tymp. tubæ; m = canal. pro tens. tymp.; v = lateral tympanic wall belonging to the pars tymp.; co = cochlea; mi = meat. audit. int.

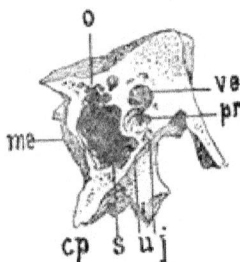

Fig. 43. — Frontal section, 3 mm. behind the previous one.—me = anterior wall of the auditory canal; s = sulcus tymp.; o = most anterior portion of the upper tympanic space; u = lower tympanic wall; ve = vestibulum; pr = promontory with the section of the first cochlear convolution.

is 1-1½ mm. The cut here passes almost exactly through the modiolus of the cochlea (co), the apex of which points directly towards the canalis pro tens. tymp. (m). Above the second convolution of the cochlea the plane of this section shows, as a rule, the hiatus canalis Fallopii.

Fourth cut (Fig. 43), 3 mm. behind the preceding. This passes through the anterior portion of the upper and middle tympanic space and through the external meatus, the anterior (me) and lower (u) walls of which are divided close to the sulcus tymp. At the medial side of the section of the tympanic cavity, the promontory, the processus cochlearis, and the canalis Fallopii are cut through close to the anterior edge of the fenestra ovalis, and with the first turn of the cochlea the vestibule (ve) also is exposed. By this transverse section the relations of the fossa jugularis (j) to the inferior wall of the tympanic cavity and to the lower surface of the pyramid become like-wise visible.

Fifth cut, 5 mm. behind the last (Fig. 44). It gives a clear topo-graphical view of the antrum mastoid. in its relation to the tympanic

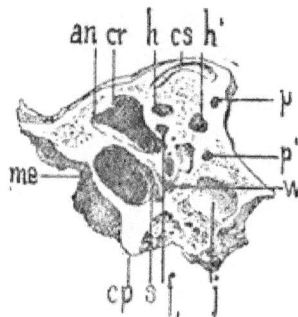

FIG. 44.—Frontal section in front of the posterior wall of the tympanic cavity.—me = meat. extern. ; s = lower segment of the sulcus tymp. ; w = posterior tympanic wall ; an = antrum mast. ; cr = crista longitud. antr. mast. ; cs = canal. semicirc. super. ; h h′ = openings of the section of the canal. semicirc. horizontal. ; p p′ = opening of the section of the canal. semicirc. post. ; f = opening of the section of the facial canal ; j = fossa jugularis.

cavity and the external meatus. As the cut falls behind the vestibule it meets all the semicircular canals and the descending portion of the canalis facialis (f).

The sections farther back concern the antrum mastoid. and the mastoid process, and will, therefore, be noticed in the chapter ' Prepara-tion of the Mastoid Process.'

Frontal cuts carried, as here described, in parallel planes, necessarily strike the external and internal meatuses in a slanting direction, as the axis of the latter forms with the long axis of the pyramid an obtuse angle opening outwards and towards the front (comp. Fig. 47). In order, therefore, to obtain a section through the temporal bone, show-ing most clearly the anatomical relations of the external meatus to the middle ear, a cut is carried in such a way as to pass through the long axis of the external meatus, and also through that of the internal,

which lies in nearly the same direction. The cut thus goes through
the axis of the external meatus, and through the middle portion of
the tympanic cavity, meets the inner wall of the latter in front of the
fenestra ovalis, divides the anterior part of the vestibule and the first
turn of the cochlea, and terminates at the internal meatus. The tem-

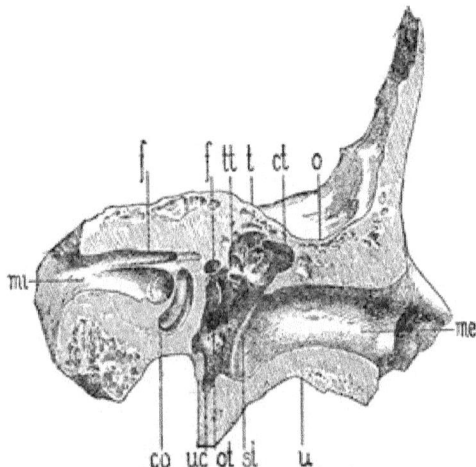

FIG. 45.—Anterior half of a temporal bone, sawn through in a frontal direction in the long
axis of the external meatus. (Right ear.)—me = meat. audit. extern. ; o = upper wall
of the meatus ; u = its lower wall ; st = sulcus tymp. ; t = tegm. tymp. ; uc = lower
tympanic wall ; ot = ost. tymp. tubæ ; tt = canal. pro tens. tymp. ; ct = fovea capit.
mallei ; mi = meat. audit. int. ; f f = canal. facialis ; co = cochlea.

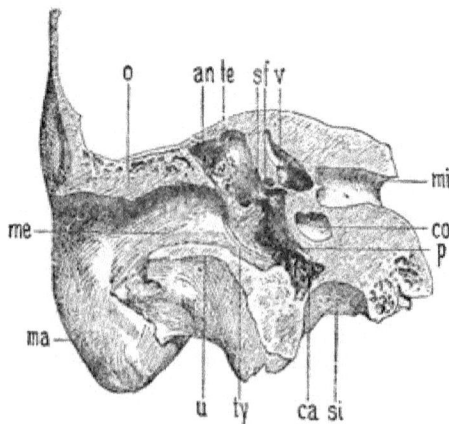

FIG. 46.—Frontal section through the temporal bone. (Posterior half.)—me = meat. audit.
ext. ; o = upper wall of the meatus ; u = its lower wall ; te = tegmen tymp. ; ca = lower
wall of the tympanic cavity ; ty = sulcus tymp. ; p = promontory ; an = antrum mast. ;
s = eminentia staped. ; f = canalis facialis ; v = vestibule ; co = cochlea ; mi = meat.
audit. int. ; ma = proc. mastoid. ; si = fossa jugularis.

poral bone, thus divided into an anterior and posterior half, shows on the
surfaces of both sections the length and thickness of the inferior (u) and

superior (o) walls of the meatus, the relative position of the latter to the middle cranial fossa; the transverse section of that part of the tympanic cavity in which the ossicula are situated, the section of the tegmen tymp. (t te), and of the inferior wall of the tympanic cavity (uc and ca), and the important relative position of this wall to the fossa jugularis. In addition to this we notice, on such sections, what is of so much importance in chronic suppurations of the middle ear, the relations of the upper tympanic space (cav. epitympan., Schwalbe), which passes into the antr. mast., to the posterior upper wall of the meatus (o), the inferior sharp edge of which is roofed over by the external portion of the cav. epitympan. This part of the latter which, as we shall see later on, serves for the reception of the head of the malleus and of the body of the incus, is sometimes the seat of an independent suppuration, leading to perforation of the membr. flaccida and to fusion of the margo tymp. with the adjoining wall of the meatus.

Frontal sections through the temporal bone are especially suited for measuring the width and height of the osseous Eustachian tube and of the tympanic cavity. The determination of the transverse diameter of the latter, in the vicinity of the tympanic membrane, can only be made on unmacerated temporal bones. The measurements noted by me on a large series of sections may be gathered from the following: Height of the osseous Eustachian tube in the middle of the canal (Fig. 40) 6-7 mm., width of tube below the canalis pro tens. tymp. $4-5\frac{1}{2}$ mm.; height of the ost. tymp. tubæ $4-5\frac{1}{2}$ mm., its width $3\frac{1}{2}-4$ mm. (according to Bezold, height 4.5, width 3.3 mm.). Anterior portion of the tympanic cavity, immediately in front of the ost. tymp. tubæ (Fig. 42), height 9-10 mm., width $3-4\frac{1}{2}$ mm. Middle portion of the tympanum, where the upper tympanic space is largest (Fig. 45), height from the floor to the roof 14-16 mm. (Bezold, on corrosion preparations, 11.73 mm.). Height at the hindmost portion of the tympanum in conformity with Von Tröltsch 15 mm. The width of the tympanic cavity immediately below the tegmen tymp. varies between 6 and 7 mm., from the incisura Rivini to the inner tympanic wall, between 5 and 6 mm., at the most posterior portion 5-6 mm. The breadth of the inferior tympanic wall shows, in the majority of temporal bones, a transverse diameter of 5 mm. The height of the posterior tympanic wall is 7-8 mm., that of the anterior wall, ascending obliquely from the floor of the tympanum to the lower edge of the ost. tymp., 2-3 mm. The distance from the floor of the tympanic cavity to the lower edge of the sulcus $2\frac{1}{2}-4$ mm. (according to Bezold, on an average, 2.71 mm.). The height of the opening leading to the antr. mast. above the posterior tympanic wall from the lower angular depression to the tegmen mast. varies between

5 and 7 mm., its width from 6-7 mm. (according to Bezold's sections from corrosion preparations, height 5.68, width - 6.69 mm.). In exceptional cases the height exceeds the width.

The superior wall of the osseous meatus (Fig. 45), which is of variable thickness, presents sometimes a nearly compact structure ; not infrequently, however, diploetic or pneumatic cellular spaces, which as a rule are situated in the neighbourhood of the superior lamella of the wall of the meatus, and which rarely are connected with the cavum tymp. itself, but communicate with the antrum mast. and the pneumatic cells lying at the posterior wall of the auditory canal. They sometimes extend from here forwards as far as the root of the zygomatic process (Kirchner). On frontal sections a longish fissure (Fig. 46) is frequently met with in the superior lamella of the upper wall of the osseous meatus ; it arises, as before stated, by the lamina tegminis tymp. overlapping the lamina interna of the pars squamosa. The section of the lower wall of the meatus passes inwards and downwards into the sharp edge of the crista petrosa. As, however, especially towards the front, the pars petrosa, and to even a greater extent the pars tymp., take a part in its formation, this crest might be more correctly called the crista tympanico-petrosa.

The great individual differences in the thickness of the tegmen tymp. and of the lower wall of the tympanic cavity can best be demonstrated on frontal sections made through a considerable number of temporal bones. In the tubal section, especially over the ost. tymp. tubæ, the tegm. tymp. is much thicker than over the upper tympanic space. Here it appears sometimes thick, compact, or cellular ; sometimes, again, as thin as paper, perforated like a sieve, or to a large extent broken through (dehiscence of the tegm. tymp., Hyrtl.*). Also the wall bounded by the fossa jugular. is sometimes thick and compact, rarely cellular, and at other times thin, bulging out bladder-like towards the tympanic cavity ; sometimes, again, dehiscent.

The third series of sections through the temporal bone is made in a horizontal direction. Three parallel cuts complete the topographical picture of the external, middle, and internal ear, as described on the former sections. The uppermost cut commences at the spina supra meatum—that is, close above the superior wall of the meatus—meets the upper tympanic space and the antr. mast., divides the superior semicircular canal, and terminates at the superior wall of the internal meatus. The second cut, lying 5-6 mm. lower, passes through the external meatus somewhat above its axis, consequently nearer to the

---

* The statement that these gaps in the tegm. tymp., arising during growth, are caused by atrophy in the bone had already been made by Henle in his ' Anatomie des Menschen,' Bd. I. 1870.

superior than the inferior wall, divides the boundary between upper
and middle tympanic space, and runs through the pyramid in a plane
lying in the middle of the internal auditory canal.   The upper aspect
of this cut (Fig. 47) shows the section of the anterior (a), superior, and
posterior (h) walls of the meatus, with the Rivinian segment (iR), the
upper tympanic space (ct) behind it, the antr. mast. (am) communicating with it, and the pneumatic cells of the mastoid process (c c')
situated in the vicinity of the sinus transversus (sl); towards the
front, in the angle formed by the pyramid and the squama, the canalis
musculo-tubarius.   On the pyramid the section shows most distinctly
the upper vestibular space (v), the cochlea (co), and the internal
meatus (mi).   The surface of the cut on the lower piece (Fig. 48)
shows in a comprehensive manner the length of the anterior, inferior,
and posterior walls of the meatus, the relative position of the latter to
the mastoid cells, the width and length of the lower tympanic space,
and the details of the middle ear and labyrinth mentioned in con-

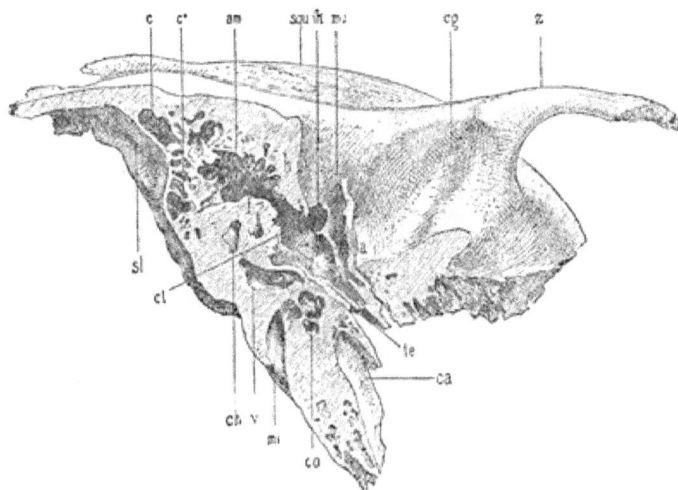

FIG. 47.—Horizontal section through the temporal bone of an adult.   (Upper half.)—squ =
squama ;   z = proc. zygomat. ;   cg = cavit. glenoidalis ;   me = meat. audit. extern. ;
ct = cavum tymp. ;   iR — incisura Rivini ;   te = Eustachian tube ;   am = antr. mast. ;
c c' = cellulæ mast. ;   mi = meat. audit. intern. ;   v = vestibule ;   ch = canal. semic.
horizontal. ;   co = cochlea ;   ca = canal. caroticus ;   sl = sinus sigmoideus.

nection with the previous series of sections, and marked in the
accompanying illustration (Fig. 48).

The length of the walls of the osseous meatus is best determined
on frontal and horizontal sections of the temporal bone; but in
making the measurements the same points at the external orifice of
the meatus and at the sulcus tymp. should always be chosen.   The
following measurements of the walls of the meatus agree generally

with the results obtained by Von Tröltsch, Bezold, and others :
Length of the superior wall of the meatus, between the incisura Rivini
and that point where the squama turns into the horizontal part,
14-16 mm. (Bezold, 14 mm.) ; length of the anterior wall, from the
anterior edge of the pars tymp. to the anterior segment of the sulcus
tymp., 15-16 mm. ; length of the inferior wall from the corresponding
edge of the pars tymp. to the inferior segment of the sulcus tymp.,
15-16 mm. ; length of the posterior wall, 14-15 mm.

With these average measurements, on the one hand, we occasionally
meet with temporal bones on which the walls of the meatus appear
longer by 1-3 mm. and more ; rarely such where the length of the
walls falls below the average. The measurements on temporal bone of
children under twelve months show the average length of the walls
of the meatus to be half of that in the adult.

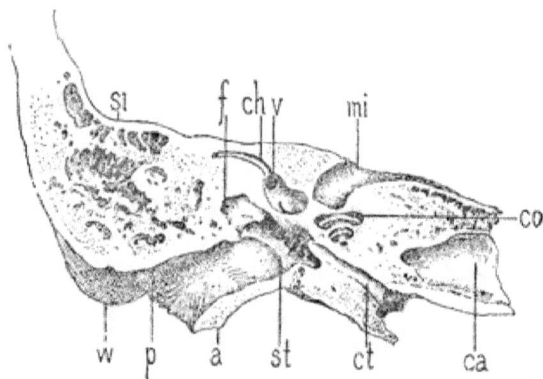

Fig. 48.—Horizontal section through the temporal bone of an adult. (Lower half.)—a =
anterior wall of the meatus ; p = its lower wall ; st = sulcus tymp. ; f = canalis facialis ;
ct = canalis pro tens. tymp. ; v = vestibule, and at its floor, the fissure leading to the
scala tymp. of the cochlea ; ch = canalis semicirc. horizontal. ; co = cochlea ; ca = canalis
caroticus ; w = proc. mast. ; st = sinus transversus.

If, on a frontal section, a perpendicular line be drawn from the
incisura Rivini to the lower wall of the meatus, the distance of this
point to the lower segment of the sulcus tymp. amounts to 4-5 mm.
(Von Tröltsch, 6 mm.). A horizontal line from the inner boundary of
the posterior to the anterior wall divides the same, 5-7 mm. from the
anterior segment of the sulcus.

The length of the inferior tympanic wall, measured between the
point where the ridged anterior wall bends towards the ost. tymp.
tubæ and the boundary of the lower and posterior walls, varies from
10-12 mm. (according to Von Tröltsch, 13 mm.; according to Bezold,
on an average, 12·73 mm.). The length of the osseous Eustachian
tube fluctuates between 10 and 12 mm.; that of the canalis pro tens.
tymp., between 12 and 14 mm.

On horizontal sections of the temporal bone made below the
Rivinian segment and the canalis pro tens. tymp., and which afford
on the upper piece a free view into the antr. mast. and cav. epitymp.,
one may frequently notice, on the lower surface of the tegmen antr.
mast., an elevation running in a longitudinal direction as far as the
tegmen tymp., which laterally and medially is flanked by small ledges
of bone standing vertically to its line of direction. This elevation,
which is not mentioned in the anatomical works I had access to, is
either solid or contains a canal, and must be regarded as the inner
boundary of the lamin. mast. int. part. squam., which is overlapped by
the tegmen mast. (comp. Fig. 27). It may therefore be designated as
crista tegm. mast. A second ledge of rather frequent occurrence, the
crista transv. tymp., first detected by Bezold in corrosion preparations
on the inferior surface of the tegmen tymp., and which describes an
arch, open downwards, on the roof of the tympanic cavity, from the
upper lamella of the proc. cochlearis to the spina tymp. major of the
Rivinian segment, gives attachment to a fold of mucous membrane
passing over to the tensor tendon. It may be called the anterior
boundary of the cavum epitymp., upon whose lateral smooth and
excavated surface (Fig. 45) lies the head of the malleus.

To obtain a clear notion of the position of the two meatuses, of the
tympanic cavities, and labyrinths, a frontal cut is made by means of
the saw through the base of a skull which has been freed from the
calvaria, passing on both sides through the axis of the external
meatus, porus acustici interni, and the posterior portion of the clivus.
The cut through each temporal bone thus corresponds to the frontal
one described on p. 42, and shown in Fig. 46. It is needless to
dwell further on the advantages which such sections through the skull
offer for estimating the relative inclination of the various cavities in
the temporal bone.

In order to make these serial sections available for study and
demonstration purposes, it is recommended to number the consecutive
pieces, and to fasten those belonging to one preparation in their
proper order with tacks on a blackboard, or to cement them on a
glass plate of the requisite size. The latter mode has the great
advantage of allowing the sections to be examined on both sides.

Another still simpler way of keeping serial sections of the temporal
bone is to bore a hole through each of the pieces, and to string them
on a piece of thin cord or wire. On preparations which are to be used
for lecturing purposes the canalis caroticus may for greater clearness
be painted over with red; the sulcus transvers., the sinus petrosus
superior et inferior with blue; and the nerve-canals with yellow
water-colour. For studying the course of the canals, as well as the

entrance and exit of vessels and nerves in the temporal bone, it is best
to make use of short and long stiff bristles, which for demonstration
may be allowed to remain.   In order that they may not fall out, they
are provided with small heads or knobs by dipping their projecting
ends into melted sealing-wax.

PREPARATION OF THE PROCESSUS STYLOIDEUS IN THE TEMPORAL BONE OF THE NEW-BORN INFANT AND OF THE ADULT.

At the lower edge of the external surface of the temporal bone of
the new-born infant there is, immediately behind the posterior crus of
the ring, between this and the foramen stylo-mast., an irregular gap,
which leads into a short osseous canal directed upwards.   This is the
sheath of the styloid process, at the lower opening of which the upper
portion of the proc. styloid. may frequently be seen if it happens to
be ossified before birth.   If at the time of birth the styloid process
was still cartilaginous, the sheath will after maceration be found
empty.

To more clearly define its position and boundary, the posterior
crus of the tympanic ring is broken off, and the lateral wall of the
sheath removed with a broad graver.   In a few cases only does its
upper portion open with a wide orifice, at the side of and below the
eminentia stapedii, into the tympanic cavity.   In most preparations,
however, the vagina proc. styl. is closed above, and its upper limit
marked by the protuberance on the posterior wall of the tympanic
cavity (eminentia styloidea), first described by me, which is situated
between the sulcus tymp. and the eminentia stapedii.

Removing the lateral wall of the sheath of the styloid process in
preparations where the upper piece of the latter is already ossified, we
find that the above-mentioned protuberance, which is visible on the
posterior tympanic wall, is caused by the club-like thickened upper
end of the proc. styl. (Fig. 49).   This club-like swelling is in some
cases very pronounced, while in others it is only slightly developed.
It is probable that from this arises the great difference in the development of the eminentia styloidea on the posterior wall of the tympanic
cavity in the new-born infant and the adult.

The position of the proc. styl., and the bulging caused by it on the
posterior tympanic wall, may also be demonstrated on successful
sections made parallel to the inner wall of the tympanic cavity,
through the eminentia styloidea, lying between the sulcus tymp. and
eminentia stapedii (Fig. 50).

In order to understand the anatomical relations of the upper piece
of the proc. styloid., which is inserted in the temporal bone, it is

advisable to expose it with the knife or scalpel on an unmacerated temporal bone of a new-born infant, after the bone has been decalcified in nitric or hydrochloric acid. On preparations where the styloid process is still quite cartilaginous, it can be completely drawn out from its sheath. Its form (Fig. 51), which I was the first to describe,

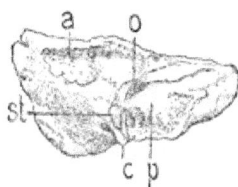

FIG. 49.—Upper end of the ossifying proc. styloid. from a new-born infant.—o = fenestr. ovalis ; p = promontory ; a = antrum mast. ; c = canal. proc. styloid. ; st = upper extremity of the proc. styloid.

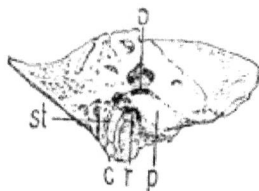

FIG. 50.—Section through the canal. styloideus in the new-born infant.—o = fenestr. ovalis ; r = fenestr. rotunda ; p = promontory ; c = canal. styloideus ; st = section of the upper ossified extremity of the proc. styloid., causing the posterior tympanic wall to bulge forward. After a preparation in my collection.

is that of a club with one or more lateral processes, enclosed, in front of the canalis Fallopii, and laterally from the eminentia stapedii, by the before-mentioned sheath. The knob-like process of its upper portion is directed forwards, and causes the posterior tympanic wall to bulge out more or less. The posterior process is lodged in a de-

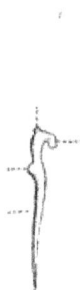

FIG. 51. — Cartilaginous styloid process from a new-born infant.

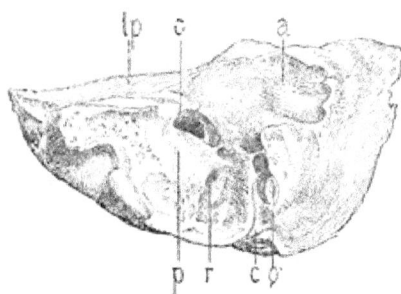

FIG. 52.—Glenoid facet on the posterior wall of the canal. styl. in the temporal bone of a new-born infant. Enlarged to double its size.—o = fenestra ovalis ; r = fenestra rotunda ; p = promontory · tp = semicanal. pro tens. tymp. ; a = antr. mast ; c = canal. styl. ; g = glenoid facet. After a preparation in my collection.

pression on the posterior wall of the vagina proc. styl.; this depression, which I was the first to describe, resembles an articular cavity, has a direction forward, and lies 5-6 mm. from the lower opening of the sheath. It is placed close to the external surface of the pars mast.

(Fig. 52), and only to be found in rare cases, because it generally coalesces with the styloid process before birth.

The preparation of the processus styl. in the adult is very difficult, because its outer compact osseous layer and the surrounding mass of the temporal bone are so completely fused that its earlier boundaries are but seldom to be recognised on sections. As, however, the styloid process frequently possesses a distinct medullary space, it is possible, on successful sections, to follow its course in the temporal bone as far as its upper termination. For this purpose either macerated or unmacerated temporal bones may be used, on which the tympanic cavity must be opened from above, in order to search for the protuberantia styl. on the posterior tympanic wall. This projection, situated laterally from the eminentia stapedii, as well as the processus styl., standing

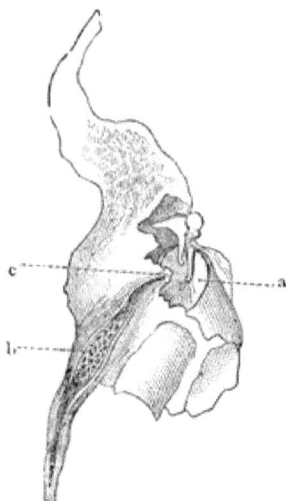

FIG. 53.—Section through the styloid process of an adult.—a = tympanic membrane ; b = medullary space of the proc. styl. ; c = its upper extremity with the protuberantia styl. on the posterior tympanic wall, first described by me.

out on the lower surface of the pyramid, serve as a guide when making the cut. The work is considerably facilitated on the macerated temporal bone if the squama and the superior wall of the meatus be first sawn away. The preparation is now fixed in the vice in such a manner that, on applying the blade of the saw to the anterior wall of the meatus, the line in which the cut is to be made falls exactly in the direction of the eminentia styl. on the posterior wall of the tympanic cavity and of the long axis of the processus styl. proper. On the surface of the section the porous medullary space of the processus styl. may in most cases be traced as far as the vicinity of the eminentia styl. The latter consists nearly always of compact bone substance.

On unmacerated preparations the medullary space of the styloid process is more clearly differentiated. The cut here takes the same direction as in the macerated temporal bone. If it be intended, however, when making the preparation, to preserve the tympanic membrane intact, after opening the tegmen tymp. the cut is carried in the longitudinal direction of the lower tympanic wall in such a manner that it passes through the eminentia styl. on the posterior wall of the cavity.

As the processus styl. in the temporal bone describes, from the lower extremity of its sheath to the eminentia styl. on the posterior tympanic wall, an inwardly convex curve, this cut divides only the upper-end piece of the styloid process. In order now to expose also the middle and lower portions of the medullary space, the osseous mass overlying it is either sawn through in the direction of the styloid process projecting on the lower wall of the pyramid, or filed down with a coarse file until the medullary space of the process (Fig. 53) comes into view.

IV.

ANATOMICAL AND PATHOLOGICO-ANATOMICAL PREPARATION OF THE ORGAN OF HEARING.

INTRODUCTION.

For the preparation of normal organs of hearing, fresh temporal bones, taken from the body soon after the post-mortem examination, are most suitable; but for anatomical study such as have been kept in Rüdinger's preservative fluid (p. 6) or spirit preparations may also be used. The employment of objects of the latter description is especially advisable where fresh temporal bones are not at all times obtainable, or where the student is tied by circumstances to certain times for his work. Those who do not wish to be interrupted in their anatomical studies should take care to have a supply of preparations of the ear kept in a preservative fluid, so that there may always be sufficient material at hand in case fresh objects should not be procurable.

It is different with the dissection of pathologico-anatomical preparations. Here it is under all circumstances advisable to begin the work of preparation on fresh objects soon after they have been removed from the body, in order to obtain as clear a representation as possible of the morbid changes in the auditory apparatus. If the preparation, for the purpose of future dissection, be kept in spirits of wine or in a preservative fluid, both colour and consistence of the object are altered in many ways, which would not admit of a correct estimate of the morbid process.

The mode of preparing in pathological cases agrees, on the whole, with that of normal preparations of the organ, though not infrequently the method is subject to certain modifications depending on the locality of the morbid changes, and on the necessity of keeping intact some parts of the object, or of entirely removing others, so as to bring the seat of the disease clearly under observation.

As the task of the pathologico-anatomical dissection of the organ of hearing lies not merely in the macroscopical representation of the morbid changes, but also in the histological investigation of the diseased tissues, it is evident that, where possible, the anatomical examination should always go hand-in-hand with the histological. This endeavour to do justice in both directions to the scientific investigation of a preparation is not infrequently beset with insurmountable obstacles. Foremost among these will be the desire of the operator to add to his collection of pathologico-anatomical preparations for teaching purposes. Here, in the interest of the instructive macroscopic representation, the histological examination will often have to be dispensed with; while in cases where, for scientific investigation of the morbid process, the importance of the histological examination is first to be considered, the preparation must be sacrificed for that purpose. The latter alternative will be the more readily chosen if the collection of macroscopical preparations contains already several similar specimens; while very rare and specially instructive pathological preparations will be selected for preservation as macroscopical objects. It should, however, be observed that not infrequently there are cases where some portion of the preparation, such as the membranous labyrinth, small portions of the external meatus, of the walls of the tympanic cavity, of the Eustachian tube, and of the mastoid process, may be taken from the specimen for histological examination without the remainder of the preparation losing anything of its value as a macroscopical object.

In the preparation of the normal organ of hearing, it is best to begin with the auricle and the external meatus, and from here to proceed with the dissection in the direction of the middle ear and the labyrinth. The manipulation is much facilitated if all the soft parts and osseous portions of the preparation, which interfere with the laying bare of the various parts of the organ, be first removed with scissors, forceps, and saw. This applies, above all, to the parotid tissue covering the lower wall of the cartilaginous meatus, to the soft parts on the squama and the lower surface of the pyramid, and to the muscles and tendinous insertions on the mastoid process. The removal of these, too tedious when done with knife and scissors, is most expeditiously effected by taking a firm hold of the preparation with a piece of cloth in the left hand; then, grasping the tissues by means of a not very sharp pair of bone forceps, twisting it round on its long axis, and giving it a vigorous pull, they are torn away. When there are no special reasons for retaining the dura mater on the object—*e.g.*, for the preparation of the aquæductus vestibuli, of the sinus transversus, etc.—the dura mater may also be stripped off with the forceps,

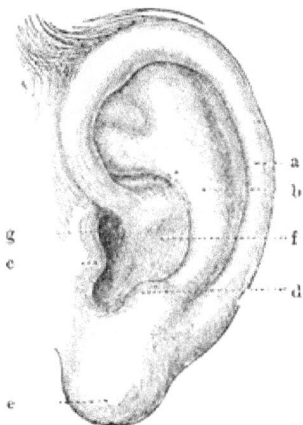

FIG. 54.—Auricle.—a = helix; b = antihelix, passing upwards into the crura bifurcata; c = tragus; d = antitragus; e = lobulus; f = concha; g = outer orifice of the meatus.

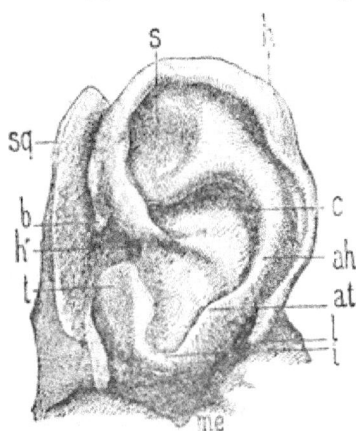

FIG. 55.—Exposed cartilage of the auricle (Left ear.)—h = helix; h″ = transition of the helix into the concha; ah = antihelix; s = fossa navicularis; t = tragus; at = antitragus; i = incisura intertragica; b = anterior pointed process of the helix (spina helicis); l = posterior pointed process of the helix (proc. helicis caudatus); me = cartilaginous meatus; sq = squama. After a preparation in my collection.

care being merely taken that the acoustic and facial nerves be not drawn out of the internal auditory meatus.

1. PREPARATION OF THE AURICLE AND CARTILAGINOUS MEATUS.

In order to become acquainted with the shape and outlines of the cartilaginous skeleton of the auricle, its integument and muscular structure, as well as the lobe, must be completely dissected off from the perichondrium. On the posterior surface of the auricle, where the skin is easily movable, this may be done without difficulty with pincette and scalpel. On the anterior surface, however, especially in the concha, at the antihelix, and in the fossa navicularis, where the integument is very firmly adherent to the perichondrium, the cutis must be detached with scalpel or scissors, great care being taken to avoid injury to the cartilage. The complete exposure of the auricular cartilage requires, therefore, considerable time.

When laid bare, the cartilage of the auricle (Fig. 55) shows, on the whole, the same depressions and elevations as it does when covered by the cutis. The most striking change of form, after dissecting off the integument, is caused by the removal of the lobe and by the spina helicis (Fig. 55, b) and the proc. helicis caudatus (l) becoming visible on the preparation.

The close connection of the cartilage of the auricle with the cartilaginous portion of the external meatus by means of a broad bridge 8-9 mm. wide, commencing at its lower portion (isthmus cartilag. auris. Schwalbe), requires, for anatomical purposes, both parts to be prepared together. This may be done by exposing the entire cartilaginous structure of the auricle and meatus; or the auricle, with its integument, is left in its natural form, and the cartilagino-membranous meatus in connection with it is dissected out from the parts surrounding it.

On a head where the auricle has been left *in situ*, the skin at the circumference of the attachment of the latter is divided by a circular incision, after which all the adjacent soft parts are dissected away. It is best to begin in the region above the auricle by removing, as completely as possible, the integument with the galea aponeurotica, the fascia temporalis and the temporal muscle from the squamous portion of the temporal bone as far as the linea temporalis and the root of the processus zygomaticus. Then the fibrous ligament (ligamentum auriculare anterius), which passes from the root of the zygomatic process to the tragus and to the anterior superior wall of the cartilaginous meatus, as well as the fibrous connections stretching thence along the superior wall of the meatus to the ligamentum auriculare post., are separated, after which the posterior superior membranous wall of the cartilaginous meatus is loosened with the narrow handle of a scalpel from its attachment to the horizontal portion of the squama (superior wall of the osseous meatus).

The skin and fasciæ over the mastoid process having been dissected off, we divide the insertion of the posterior surface of the auricle at the external surface of the latter, as well as the ligamentum auriculare post. passing from the anterior part of the external surface of the mastoid process to the auricle, and then detach with the handle of a scalpel the loose tissue-connection of the posterior wall of the cartilagino-membranous meatus with the anterior surface of the mastoid process and the outermost section of the posterior wall of the osseous meatus. This done, the laying bare of the anterior and lower walls of the cartilaginous meatus is proceeded with by so far removing with scalpel and pincette the surrounding soft parts, consisting in a great measure of parotis and adipose tissue, that the inner tapering end of the lower cartilaginous wall comes clearly into view.

The cartilaginous meatus, when thus exposed, is found united by an easily movable intervening mass of connective tissue to the exterior margin of the tympanic portion of the osseous meatus, which is covered with dense elastic fibrous tissue. The posterior superior membranous wall of the meatus, on the other hand, passes uninterruptedly into the subcutaneous connective tissue of the posterior superior wall, giving off, as we shall see, fibrous filaments to the membrana tympani.

For the study of the topographical position of the cartilaginous and osseous portions of the auditory canal, it is recommended to make a series of preparations, some showing the cartilaginous meatus in its natural position, with its intact walls; while others display, partly on

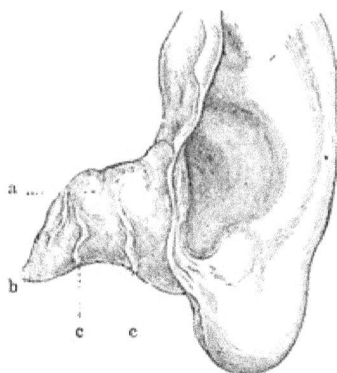

FIG. 56.—Auricle and cartilaginous meatus. (Left ear.)—a = cartilaginous meatus; b = inner pointed extremity of the cartilaginous meatus; c c = incisuræ Santorinianæ.

horizontal, partly on frontal sections (comp. the chapter, 'Making Topographical Sections of the Organ of Hearing for Instruction Purposes'), the course of the external auditory meatus.

To obtain the auricle and cartilaginous meatus as a separate anatomical preparation, the distinctly marked bridge of connective tissue between the cartilaginous meatus and the pars tymp. of the osseous meatus, as well as the posterior superior membranous wall of the pars cartilaginea, are cut through.

The cartilaginous meatus after separation appears, as seen in front (Fig. 56), as a channel gradually narrowing inwards, the lower inner extremity of which terminates in a rounded point (b), which slightly projects below the inferior wall of the osseous meatus.*

* This tapering extremity of the cartilaginous meatus, described by Schwalbe (Anatomie des Ohres, 1886) as process. triangularis, was described and illustrated in the first edition of my 'Text-book,' 1878, p. 6.

The cartilaginous portion of the external meatus is traversed vertically to its long axis by two fissures (c c, incisur. Santorini), which are occupied by connective tissue, and the anterior of which (incisur. major) lies within the region of the cartilage of the tragus. Parotid abscesses not infrequently break through these fissures into the external meatus. In the cartilage of the auricle, the thickness of which varies at different points from 1-3 mm., there are also inconstant fissures and openings (Sömmering, Schwalbe) closed by connective tissue, which, according to Parreidt ('Dissertatio Inauguralis,' 1864), allow of the anastomosis of the bloodvessels between the anterior and posterior surfaces of the auricle.

To demonstrate the limits of that region up to which the

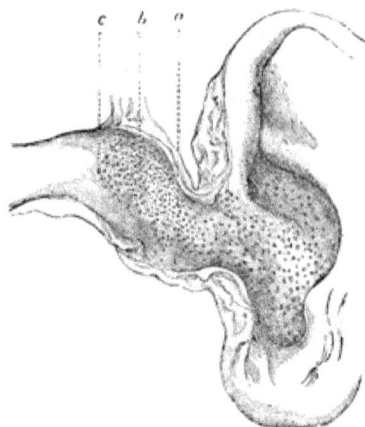

FIG. 57.—Posterior wall of the cartilaginous and osseous meatus.—a = orifices of glands in the cartilaginous portion ; b = boundary between the cartilaginous and osseous meatus ; c = pointed extremity of the triangular space, occupied by the orifices of the glands, which projects into the osseous meatus.

ceruminous and sebaceous follicles extend in the external meatus, the anterior wall of the cartilaginous meatus is cut away with scissors, as seen in the accompanying illustration (Fig. 57), part of the anterior lower wall of the osseous portion being broken off with bone forceps. It will be seen from such preparations that the glandular elements do not occur, as was formerly supposed, in the cartilaginous meatus alone, but that they extend, as first proved by Von Tröltsch, in the form of a triangular wedge (b c) into the osseous meatus.

The cartilage of the ear, as a preparation, keeps its form best in spirit ; while, if dried, it shrinks so as to become unrecognisable. To obtain useful dry preparations of the auricular cartilage, or of the auricle covered with its integument, the object, either fresh or dehydrated in alcohol, should be placed in glycerine and allowed to remain from eight to fourteen days. The saturated preparation, fixed with pins on a small wooden slip, and protected from dust, retains for years its shape and pliancy, especially if from time to time it be brushed over with glycerine.

Preparations of congenital or acquired deformities of the auricle : neoplasms of this part, epithelioma, sarcoma, cysts, auricular appendages, etc., must for demonstration purposes be preserved in alcohol. Where the new growth is also to be subjected to microscopical examination, small portions of it should be cut off from the preparation, care being taken not to spoil the appearance of the object.

PREPARATION OF THE MUSCLES OF THE EXTERNAL EAR.

The representation of the muscular apparatus of the auricle, which, according to Ruge's* investigations, must be considered as the remnant of the platysma myoides, may, with sufficient care, be successfully made on fresh preparations ; but it is more to the purpose to follow Ruge's method, which is to place the object for some days in a weak solution of chromic acid (1·2000), and then to harden it in alcohol. The work of preparing is thus facilitated, as the muscular fasciculi, by their deeper colour, appear sharply defined against the subcutaneous connective tissue and fasciæ.

The muscular apparatus of the auricle is divided into two groups. The first of these, consisting of three principal muscles, effects the movement of the entire auricle. The m. attollens s. levator auriculæ, the most strongly developed, arises, with its fan-shaped radiating bundles, at the galea aponeurotica of the temporal region, and is attached, with its fasciculi converging downwards, to the convex surface of the auricle. The m. attrahens auriculæ, which draws the auricle somewhat forwards and upwards, arises in front of the ear from the root of the proc. zygomaticus, and has its point of attachment at the crista helicis. The m. retrahens auriculæ arises with several distinct fasciculi from the mastoid process, and is inserted at the posterior convex surface of the concha. Having ascertained theoretically the position of this group of muscles, there will be no difficulty in finding them during the preparation, if care be taken, when dissecting off the skin, that only the most superficial layers of the subcutaneous connective tissue are removed with the outer integument.

The preparation of the second group of muscles, which have their origin and points of attachment on the auricle itself, and which, even if strongly developed, can only effect a slight change in the shape of

* Untersuchungen über die Gesichtsmusculatur der Primaten, Leipzig, 1887.

the auricle, presents far greater difficulties, because the various pale muscular bundles are for the most part very faintly developed, and, in dissecting off the cutis, may easily be removed along with it. The bundles of the m. helicis major pass from the anterior edge of the helix to the spina helicis; those of the m. helicis minor run on the lower portion of the helix, which projects into the concha. The m. tragicus passes, with its vertical fibres, on the outer surface of the tragus; the m. antitragicus in the direction of the lower extremity of the antihelix to the antitragus. On the posterior surface of the auricle the m. transversus auriculæ, made up of several transverse bundles corresponding to the convexity of the concha, is easily found, because the skin, being movable, can be readily dissected off from the underlying structure.

2. OPENING OF THE EXTERNAL MEATUS. LAYING BARE OF THE EXTERNAL SURFACE OF THE MEMBRANA TYMPANI.

In the anatomical dissection of the normal organ of hearing, it is advisable, considering the topographical relations of the external meatus, to retain the cartilagino-membranous portion of it *in situ*; and, on opening the canal, to remove only the anterior cartilaginous wall. In pathologico-anatomical dissections, on the other hand, where there is seldom occasion to retain the auricle and the external portion of the cartilaginous meatus on the preparation, the examination of that part of the cartilagino-membranous meatus still connected with the object will only become necessary if any morbid changes should be noticeable in it. Where this is not the case, the work of preparing is materially simplified if, before opening the osseous meatus, the cartilagino-membranous portion be entirely removed.

(A) Opening of the External Meatus in the Anatomical Dissection of the Normal Organ of Hearing.

The laying bare of the external meatus and of the external surface of the tympanic membrane is, after removal of its anterior cartilaginous wall, most easily effected by taking away the anterior and part of the inferior osseous wall. The instrument most suitable for this purpose is the bone forceps (Fig. 10) shown on page 3, with which the usually thin anterior osseous wall may readily be broken away bit by bit. Only in rare cases does this wall attain such thickness that it can only in part be broken down by the forceps, and, in order to complete the work more speedily, chisel and saw have to be employed.

The nearer we come to the tympanic membrane when opening the osseous meatus, the more cautiously must forceps or chisel be handled, so as not to injure the membrane or break its frame. It is therefore advisable, when approaching the latter, to remove the lamellæ of the anterior wall of the meatus in very small pieces, using the smallest forceps only. Any projecting bits of bone which cannot be grasped with the forceps, and which intercept the view into the tympanic cavity, are to be carefully scraped off by means of a narrow, sharp hand-chisel or graver (Fig. 12). This part of the work requires, however, a steady hand and great caution, as a slight slip might lead to the destruction of the tympanic membrane.

Projecting portions of the thick superior or inferior wall of the meatus, which prevent the examination of the membrana tympani, are chipped off, but in layers only, with chisel and hammer, or cut away with the saw. The latter manipulation is, with some practice,

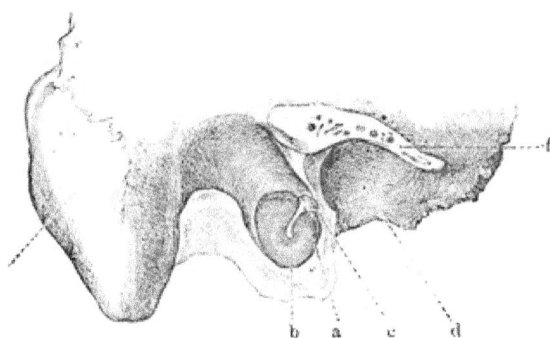

FIG. 58.—Outer surface of the tympanic membrane, natural size. (Right ear.)—a = short process of the malleus; b = lower end of the manubrium (umbo); c = membrana flaccida Shrapnelli; d = cavitas glenoidalis; e = mastoid process; f = section of the zygomatic process.

to be preferred, because it is more expeditious and less risky. When operating with the bone forceps, the object, partly wrapped in a piece of linen, is firmly held in the left hand; but if chisel or saw be used, it should be fixed in the vice. After the removal, in the regular way, of the anterior inferior wall of the osseous meatus and a portion of the superior wall as far as the vicinity of the sulcus tymp., the external surface of the tympanic membrane appears completely exposed, as shown in Fig. 58, and affords a clear view of its details: the short process (a), the handle of the malleus (b), Shrapnell's membrane (c), and the curvatures of the tympanic membrane. Thick flakes of epidermis, which not infrequently, even in preparations of the normal organ, cover the membrane wholly or partially, and which adhere mostly to the cutis, interfering thus with the normal appearance of the membrane, must be carefully brushed off, but not removed with the pincette, as this would easily cause rents in the membrane.

(B) Opening of the External Meatus and Exposure of the Outer Surface of the Membrana Tympani in Pathological Dissections.

In pathological dissections, before laying bare the external meatus and the outer surface of the tympanic membrane, the auditory canal must be examined with the speculum and reflecting mirror, in order to remove, by syringing, any ceruminous and epidermic masses or purulent secretion which may have accumulated there. Every pathological dissection requires, moreover, previous to the opening of the external meatus, the following experiments to be made, which are of importance for estimating the middle ear processes :

1. To obtain information as to the permeability of the Eustachian tube, the presence of secretions in the tympanic cavity, and the condition of the tympanic membrane, air is forced into the tympanum by means of a canula provided with an india-rubber bag, and introduced into the cartilaginous tube, the sounds arising therefrom being examined with an auscultation tube inserted into the cartilaginous meatus.

2. Then the tympanic membrane is illuminated, in order to observe the mobility of its various parts while air is being forced into the tympanic cavity. It will then be seen whether the membrane is totally or partially thinned or thickened, whether extensive or partial adhesions exist between it and the inner wall of the tympanic cavity, or if it be perforated. Non-adherent, especially thinned portions of the membrane and isolated cicatrices will bulge strongly outwards, frequently in the form of bladders; while adherent spots remain immovable, or show only traces of mobility. When the tympanic membrane is perforated, air rushes with an audible noise through the perforation, carrying at times secretion along with it from the tympanum into the external meatus.

3. Where, in pathological dissections, rigidity or anchylosis of the ossicula has been supposed to be the cause of the disturbance of hearing during life, the opening of the tympanic cavity, which is to be described further on, must precede the laying bare of the outer surface of the tympanic membrane. If, after removal of the tegmen tymp., the malleo-incudal articulation is exposed and does not appear enveloped by masses of connective tissue, the mobility of this joint can be most easily ascertained by fitting air-tight into the external orifice of the ear, or into the lumen of the detached cartilaginous meatus, an olive-shaped nozzle, with a piece of india-rubber tube ¼ metre long fixed to it, for the purpose of alternately rarefying and condensing the air in the auditory canal. With normal mobility of the malleo-incudal joint, it will be possible with the naked eye, but

better still with a lens, to notice any displacement of the surfaces of the articulation; while in anchylosis of the joint, the body of the malleus and incus, as a whole, will prove to be movable only to a slight degree. If, at the same time, the short process of the incus has become firmly fixed in its saddle-shaped depression, or if its long process should be found adherent to the posterior tympanic wall, the small bone will not be movable at all. Still, the malleus may show a certain degree of mobility while, in anchylosis of the incus alone, the malleo-incudal articulation is not implicated, as I observed to be the case on two preparations taken from deaf-mutes.

In those not very rare cases, where hearing was disturbed to a high degree in consequence of anchylosis of the stapes in the fenestra ovalis, its presence cannot be proved in the manner here described by inspection from above, because, even with normal mobility of the stapes, it will be found impossible, although a magnifying lens be used, to perceive any locomotion in the incudo-stapedial connection. Anchylosis of the stapes, however, may be demonstrated by a simple method, first pointed out by me. It is as follows: The superior osseous semicircular canal, recognisable by the eminentia arcuata on the upper surface of the pyramid, is opened with chisel or file, and then filled with a drop of fluid so as to let it bulge out above the level of the canal. If now the air in the external meatus be condensed and rarefied, or if slight pressure be exercised with a probe upon the short process of the malleus, or upon the body of the malleus and incus, it will be observed that—provided the malleus and incus are not fixed—there is, with a movable stapes, a distinct locomotion of the speck of light on the drop of fluid; while, in anchylosis of the stapes, this light-reflex will remain stationary. Introducing the tube of a manometer containing coloured fluid air-tight into the superior semicircular canal, for the purpose of demonstrating the fluctuations of the labyrinthine fluid, is more suited for physiological experiments than for demonstrating anchylosis of the stapes.

The proceeding in opening the external meatus in pathological cases is, on the whole, the same as we have just described for the normal organ of hearing. In rare cases only, where, previous to opening the external auditory canal, it had been proved by ocular inspection that there existed pathological changes on the anterior wall of the cartilaginous or osseous meatus, such as breaking of a parotid abscess through one of the Santorinian fissures, caries or fracture of the anterior wall of the osseous meatus, formation of cholesteatoma in the same, etc., it will be necessary, for the purpose of plainly showing the pathological changes, to remove either the inferior or the superior wall of the meatus. If there be any changes on the anterior wall, it is, under certain circumstances, even necessary to remove part of the posterior wall and the mastoid process, in order to obtain a clearer view into the external meatus.

In private pathological dissections, where the auricle must not be removed, and where the cartilaginous meatus has been cut through about its middle, the remaining portion of the cartilagino-membranous canal is, before opening the osseous meatus, removed in the following simple manner :

The remains of the temporal muscle still attached to the squama, the fasc. temporal., and galea aponeurot. are quickly drawn off with the periosteum from above downwards as far as the entrance into the osseous meatus. This is best done by firmly laying hold of the soft parts close to the bone with the pointed bone forceps, and turning the latter on its long axis. The soft structures having been removed as far as the upper periphery of the osseous meatus, that part of the cartilaginous meatus which is held by the ligamentum auriculare ant. et post. is then also grasped with the forceps, and by a single steady twist completely separated from the osseous meatus. If this be done too quickly, the cutis of the superior wall of the meatus will be torn out up to the tympanic membrane and Shrapnell's membrane, or the posterior superior segment of the tympanic membrane injured.

In those pathological dissections where, during life, a chronic middle ear suppuration with symptoms of caries of the temporal bone had existed, the opening of the external meatus is followed by a minute examination of its walls. After thoroughly washing away the secretion, we notice in carious processes or in cholesteatomatous formation in the temporal bone, most frequently on the posterior superior wall of the meatus, jagged fistulous openings which, either free or covered with granulation tissue, communicate with the mastoid cells or the antrum, or else are in connection with fistulous passages opening out at various spots in the neighbourhood of the ear. The length and direction of these fistulous passages are to be ascertained with fine elastic probes. Should the examination, on account of considerable tortuosity of the fistulous channels, fail to give any result, it is often possible, by injecting them, to demonstrate their outlet in the external meatus.

The external surface of the tympanic membrane, being exposed, becomes now the subject of thorough examination in every direction. For this purpose layers of macerated epidermis, inspissated secretion, etc., which are frequently present, are removed with a camel's-hair brush dipped in water, or washed away with the ball syringe, an examination being afterwards made to ascertain whether the tympanic membrane be normal or appear morbidly changed. In the latter case it should be examined—best with the aid of a lens—whether the cuticular layer of the membrana tympani be smooth or loosened; whether small ulcers, papillary excrescences, granulations, or polypoid proliferations can be proved to exist on it; whether it be normal in its curvature or strongly retracted, and, funnel-like, indrawn, with marked prominence of the proc. brevis and of the posterior fold. If the membrane be perforated, particular note should be taken of the exact spot, size, and form of the opening, and an examination made to see whether its edges are free or in contact with the inner wall of the tympanic cavity; or, finally, whether they are united with it. Further, it must be ascertained if the opening of the perforation be blocked with masses of epidermis or granulations, which latter not infrequently find their way through the gap into the external meatus. By direct inspection it is now also possible to see more plainly any atrophied portions and cicatrices observed during the propulsion of air into the tympanic cavity, also bubble-like bulgings, as well as the extent of any casual adhesions.

Should there be any extensive defects of the tympanic membrane, the morbid changes of the inner, lower, and posterior walls may also be observed through the gap. Attention ought first to be directed to the condition of the mucous membrane, to ascertain if it be loosened, granulating, hypertrophied from polypous growths, or shrunken and sclerosed; whether the handle of the malleus appear abnormally inclined inwards and adherent to the promontorial wall, or be carious, and partially or totally destroyed; whether the incudo-stapedial connection be intact, or the long crus of the incus defective. Finally, should there be complete destruction of the membrane, the changes in the niches of the fenestra ovalis and the fenestra rotunda, and in the vicinity of the ost. tymp. tubæ, can now also be seen. Inspissated secretions and epidermis masses lodging in the tympanic cavity must, of course, first be thoroughly washed out with the ball syringe.

Where, during life, a purulent process had existed in the middle ear, note should further be taken, in laying bare the external surface of the tympanic membrane, of the condition of Shrapnell's membrane, which is situated above the short process. If it be perforated, generally a small fistula-like opening will be found, through which, on introducing a probe, the smooth or carious neck of the malleus can be reached. Not infrequently, however, cholesteatomatous masses, or polypi connected with the neck of the malleus, find their way through the gap into the external meatus (Eugen Morpurgo), proliferating over part of the tympanic membrane, and allowing its position to be ascertained only by careful probing. If there be extensive gaps above the Rivinian segment, arising through fusion of the margo tymp. above Shrapnell's membrane, it is very easy, after washing out the secretion from the upper tympanic space, to discover whether the malleo-incudal connection be intact or carious, and, by careful probing, to make out the condition of the walls of the upper tympanic space, and sometimes also of the antrum mastoid eum.

3. OPENING OF THE TYMPANIC CAVITY BY THE REMOVAL OF THE TEGMEN TYMPANI.

The external surface of the tympanic membrane having been exposed, the opening of the tympanic cavity is proceeded with. This is most expeditiously effected by cutting away the roof of the tympanum (tegmen tymp.). If, after stripping off the dura mater from the preparation, the tegmen tymp. be found firm and compact, the object is fixed in the vice to remove, with a narrow straight chisel, that part of the tegmen tymp. which lies to the side of the eminentia arcuata of the pyramid, formed by the superior semicircular canal. But in order to avoid, during the proceeding, dislocating the ossicula, which are situated immediately below the roof of the tympanic cavity, it is well to keep at first so far behind that the tegmen antri mastoid. be first chiselled away, and the tegmen tympani proper removed afterwards. Should, however, the roof of the tympanum prove thin, transparent, and in some places dehiscent, it is possible, while firmly grasping the preparation in a piece of cloth in the left hand, to break away with the pointed bone-forceps the roof of the tympanic cavity in a very short time, and thus expose the contents.

To gain an unobstructed view into the tympanic cavity, the tegmen tymp. is removed in its entire breadth, and the opening of the cavum tymp. extended so far backwards and forwards as to expose simultaneously the antrum mastoid. and the osseous Eustachian tube, after previously drawing out the m. tensor tymp. from its osseous canal.

When the tympanic cavity has been opened from above (Fig. 59), the malleo-incudal articulation (ha) in the upper tympanic space comes first into view. It divides the upper tympanic space (cavum epitympanicum, Schwalbe) into an inner and an outer portion. The anatomical relations of the latter will, on account of their great pathological importance, be discussed more fully later on.

In front of the malleus, at the boundary between the middle and upper tympanic space, the tendon of the m. tensor tymp. (s) passes transversely through the tympanic cavity. In a line with this, farther backwards, in the middle tympanic space, the incudo-stapedial connection, with the tendon of the stapedius, is visible. But while the capitulum of the stapes is clearly to be seen, its crura, when being searched for in the cavity from above, are almost entirely hidden by the upper wall of the niche of the pelvis ovalis and the prominating facial canal. With a favourable light the upper segment of the inner surface of the tympanic membrane, with the handle of the malleus, part of the inner tympanic wall, the ridged floor of the tympanic cavity, and the transition of the tympanic walls into the osseous Eustachian tube, may also be seen through the cavum epitympanicum.

Besides the details here enumerated, the tympanic cavity, when opened, displays partly constant, partly inconstant, folds of mucous membrane stretched in it. Concerning these, we refer to the description, to be given hereafter, of the preparation of the ligamentous apparatus of the malleus and incus. For the present, attention need

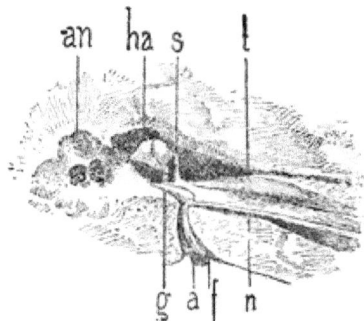

FIG. 59.—View of the tympanic cavity after removal of the tegmen tymp. (Right ear.) —ha = malleo-incudal articulation : t = musc. tens. tymp. ; s = tendon of the musc. tens. tymp. passing across the tympanum ; f = nerv. facialis ; g = genu nervi facialis ; n = nerv. petros. superf. major ; a = nerv. acusticus ; an = antrum mast. After a preparation in my collection.

only be called to the fact that these inconstant delicate folds and filaments, stretched between the tympanic membrane, ossicles, and walls of the tympanum, are, as I was the first to point out, to be looked upon as the residue of the gelatinous connective tissue which during foetal life fills up the middle ear. Their occurrence is, in so far, of importance, as these normal bridges of mucous membrane, which may easily be taken for pathological connective-tissue new growths, are without doubt, in inflammatory middle ear processes, the cause of adhesive strands of connective tissue.

EXAMINATION OF THE TYMPANIC CAVITY IN PATHOLOGICAL DISSECTIONS.

In pathological dissections, after the removal of the tegmen tymp., note is to be taken : 1. Whether serum, mucus, pus, or masses of epidermis have accumulated in the tympanic cavity, and what changes its lining and the mucous membrane covering of the ossicula show after the secretion has been removed.* 2. Whether the tympanic space be narrowed by swelling of the mucous membrane, or filled partially or entirely by granulation tissue, polypous proliferations, or organized connective tissue. The latter is most frequently found in the upper tympanic space immediately below the tegmen tymp., as a reddish-gray or yellowish-red succulent or dense connective tissue, which completely envelops the malleus and incus, and often also the stapes. In these cases the middle and lower parts of the tympanum are often found free from any new formation of connective tissue. Complete filling up and obliteration of the tympanic cavity by newly-formed masses of connective tissue (Toynbee, Von Tröltsch, Politzer) are less frequently observed. 3. Whether newly-formed strands and bridges of connective tissue appear stretched between the tympanic membrane, the ossicula, and the walls of the tympanic cavity; in what manner the position of the ossicles is altered thereby; and what changes, in consequence of such adhesion bridges, may exist on the external surface of the tympanic membrane. 4. Whether the ossicula, upon careful sounding, show normal mobility, whether they are rigid or anchylosed with the tympanic walls; finally, whether malleus and incus have been dislocated, extruded, or partially or entirely destroyed by caries and necrosis.

Frequently it is sufficient, in order to ascertain the pathological changes in the tympanum, to simply remove the tegmen tymp. With some pathological appearances, however, in the tympanum, such as adhesion of the head of the malleus, or anchylosis of the body of the malleus and incus with the upper tympanic wall, or proliferations of the mucous membrane filling up the upper tympanic space, and enveloping the body of the malleus and incus, where, consequently, the examination of the tympanum from above is rendered very difficult, it is often necessary—after previously opening and minutely examining the osseous and cartilaginous Eustachian tube—to expose the tympanic cavity from the front and from below, so as to obtain a view into the middle and lower tympanic space.

According to the seat of the pathological changes, it sometimes suffices in such dissections to break off, with the bone forceps, the external wall of the osseous Eustachian tube and of the tympanum up to the annulus tymp., while in other cases the entire Eustachian

---

* Fluid and semi-fluid secretions must be removed from the tympanic cavity by suction with a pipette; epidermis masses should be taken out, partly with the probe, and partly with the ball-syringe. A strong stream of water for cleansing the cavity from secretion is to be avoided, because thereby the mucous membrane is denuded of its epithelium, and sequestrated portions of the ossicles may be washed away.

tube, along with the anterior extremity of the pyramid and the inferior tympanic wall as far as the sulcus for the tympanic membrane, must be removed with saw, chisel and forceps, so as to bring clearly into view any coalescences between the handle of the malleus and the inner tympanic wall, or adhesions between the membrana tympani, the long process of the incus and the promontory, etc. In adhesions, and especially pathological strand-like formations in the middle tympanic space—if the preparation is to be kept for demonstration purposes—the space must be so cleared by removing any portions of bone which might obstruct the view into the cavum tymp., that the morbid changes may become clearly visible.

With some preparations this can only be done by sawing away, in a horizontal direction, the upper third of the pyramid up to the height of the tendon of the tensor tympani; while with others, where the changes are located in the middle and lower tympanic space, it must be done by horizontally separating the lower third of the pyramid as far as the immediate vicinity of the fenestra rotunda, the niche of which then also comes into view. In a large number of the preparations in my collection, the pathological changes in the tympanic cavity have been laid bare after this method.

It is evident that even in preparations with instructive pathological strand-formations in the tympanic cavity, the cutting through of these, and the separation of the outer tympanic wall, along with the membrana tympani from the pyramid, cannot be avoided if, in order to render clear the morbid appearances, a minute examination of the niches of the oval and round windows, or of the labyrinth, becomes necessary. In those pathological dissections, however, where, after opening the tympanic cavity, neither masses of connective tissue are found in it, nor adhesions met with between the tympanic membrane and the inner tympanic wall, the membrana tympani, together with the malleus and incus, must, under all circumstances, be separated from the pyramid, so as to submit the changes in the niches of the fenestra ovalis and the fenestra rotunda to a thorough examination.

The proposal of Voltolini (l.c.) not to open the tegmen tymp. in pathological dissections, but to examine the changes in the tympanic cavity from the external meatus, after cutting out the tympanic membrane and the handle of the malleus, has, for good reasons, not found favour in any quarter. This proceeding not only leads to the destruction of the morbid changes on the tympanic membrane and of the frequently-occurring bridges of adhesion between the tympanic membrane and the inner tympanic wall, but it excludes also a more minute examination of the pathological products in the upper tympanic space, and in the lateral recesses of the cavum tymp., since these parts, even after complete removal of the tympanic membrane, can only be imperfectly inspected.

4. SEPARATION OF THE MEMBRANA TYMPANI, WITH THE MALLEUS AND INCUS, FROM THE PYRAMID OF THE PETROUS BONE.

Where, in normal or pathological preparations, the inner surface of the tympanic membrane on the one side, and the details on the inner tympanic wall on the other side, are to be brought into view, it is necessary to have recourse to a method of separating the tympanic membrane from the pyramid, whereby dislocation of the ossicles may be avoided with certainty.

The middle ear may be divided sagitally, either by carrying a cut simultaneously through the tympanic cavity, the mastoid process, and the Eustachian tube in its entire length, or without having regard to the latter by simply separating the pyramid from the pars tymp. et squamosa.   In pathological dissections the separation of the pyramid must always be preceded by the examination of the tubal canal (see the chapter ' Preparation of the Eustachian Tube ').

As regards the separation of the pyramid from the pars tymp. et squamosa, it is to be observed that after removal of the tegmen tymp. the tendon of the tensor tymp. (Fig. 59 s), passing in front of the malleus across the tympanic cavity, is divided with a narrow small knife, and the incudo-stapedial connection also carefully severed. The separation of the tympanic membrane, together with the malleus and incus, from the pyramid, may now be proceeded with in the following manner: The preparation, with the opened tympanum directed exactly upwards, having been fixed in the vice, is sawn through along the floor of the osseous Eustachian tube so far towards the tympanic cavity, until the vicinity of the incudo-stapedial articulation is reached between the tympanic membrane and the inner tympanic wall.

Now, in order to guide the blade of the saw between the tympanic membrane, malleus and incus on the one side, and the capitulum of the stapes on the other side, without dislocating or injuring any of the ossicles during the operation, the tympanic membrane, with the handle of the malleus, and the long process of the incus, are pushed so far outwards with a probe held in the left hand that the narrow saw blade may, with due care, be carried through between the crus of the incus and the head of the stapes towards the posterior tympanic wall. This having been reached, it is possible without any further caution being required, to complete the cut through the mastoid cells and the sinus sigmoideus.

As, however, with insufficient practice, the incus may be dis-located, or the crura of the stapes broken off, it is advisable, for the pro-tection of the membrana tympani and the ossicula, to use a thin metal plate bent to the shape of the letter U, measuring 3 cm. by 1 cm., its anterior and lower edges being curved, to correspond with the posterior and inferior tympanic walls. The distance between the two branches of the plate, when bent towards each other, amounts to ¼ mm., which gives free play to the saw while passing to and fro between them. Having penetrated with the blade of the saw, in the manner previously described, up to the vicinity of the membrana tympani, the metal plate is fixed on the posterior edge of the saw blade, then carefully pushed in between the crus of the incus and the capitulum of the stapes in such a way that the anterior curved edges of the plate touch the posterior and lower walls of the tympanic cavity.

Now, pressing down the metal plate with the index-finger of the left hand, the lower and posterior tympanic walls and the inner portion of the pars mastoid. may be sawn through without injuring the tympanic structures.

Another method of separating the external tympanic wall, with the tympanic membrane, from the pyramid—which, however, should only be employed after sufficient practice—consists in removing with the forceps the anterior and lower tympanic walls, after having pre-viously cut through the tendon of the tensor tympani and the incudo-stapedial articulation. The preparation is then fixed in its normal position in the vice, and by means of a chisel 1 cm. broad, and applied vertically 1 mm. from the inner side of the short process of the incus, the compact posterior tympanic wall, as well as the posterior wall of the pyramid bordering on the anterior edge of the sinus sigmoid., are severed with a few blows of the chisel.

With greater certainty, and equally expeditiously, the task will be accomplished with preparations of this kind if, instead of a chisel, a strong pair of forceps be used, with which the posterior pyramidal wall between the exposed antrum mastoid. and the bare floor of the tympanic cavity is broken off close in front of the sinus sigmoid., after which strong pressure applied to the apex of the pyramid is sufficient to detach the latter entirely from the pars squamosa et tymp.

In the case of children the separation of the membrana tympani from the pars petrosa may be most readily effected with small pointed bone forceps. After removing, as in the former methods, the tegmen tymp., and cutting through the tensor tendon, as well as the incudo-stapedial articulation, the lower wall of the osseous Eustachian tube, and the thin inferior tympanic wall as far as the foramen stylomast., are removed with the forceps, whereby the anterior and inferior

segments of the tympanic ring are laid bare. Now, the osseous substance of the pars mastoid, adjoining the posterior crus of the annulus
is also divided with the small bone forceps, but only so far as to leave
the point of the annular crus in connection with the pars squamosa.
If, then, the sutura mast. squamosa be also broken through, the
tympanic membrane (Fig. 60), framed in by the annulus and the
lower portion of the pars squamosa, will fall away from the pars
petrosa.

The method proposed by Lucæ* of opening the tympanic cavity
and separating the pyramid, a short description of which here follows,
differs in some respects from the one given above :

' Having examined the auscultatory appearances with catheter and
otoscope, I divide the entire crista petrosa into three equal portions,
and carry, 2 mm. behind, where the posterior third commences, a cut
across the whole of the temporal bone in a plane which divides the

FIG. 60.—Outer surface of the tympanic membrane with the annulus tymp. and the lower
portion of the squama of a new-born infant.   After a photograph taken by Dr. Hrubesch.

crista petrosa exactly at a right angle.   By this cut, which falls
behind the lower portion of the Fallopian canal, and behind the vertex
of the posterior semicircular canal, the greater part of the mastoid
process is separated from the pyramid, and access to the tympanic
cavity gained from behind.   It must be admitted that there is one
drawback in this : sometimes sawdust will get into the tympanic
cavity during the operation, but it can easily be removed with a
camel's-hair brush.   This cut, moreover, removes almost the whole of
the cartilaginous and part of the osseous meatus ; so that, with the
light falling directly into the tympanum, it is possible to obtain a
view of the tympanic membrane, for the more thorough examination
of which the anterior wall of the remainder of the osseous meatus is
entirely removed by means of the gouge-forceps.   It is now very easy
to gradually remove, with the scissors, the roof of the tympanic cavity

* Virchow's Archiv., Bd. 29.

from behind, without running the risk of injuring the ossicles, the mostly visible short process of the incus serving as a guide. The Eustachian tube, previously carefully examined with a probe, is now completely opened with scissors and a small chisel, when a sufficiently clear view into the tympanum will be obtained from above. This method of removing the tegmen tymp., recommended by Toynbee and Tröltsch, I consider absolutely necessary in order to ascertain accurately the presence of any casual adhesions, etc. After severing the tendon of the tensor tymp., and disarticulating the connection between incus and stapes, I detach the lower half of the tympanic membrane with a sharp knife, and then chip off, with a stout chisel, the remainder of the osseous mass of the squamous and mastoid portions, with the upper half of the tympanic membrane in connection. If the soft parts be then cut through, the separation of the

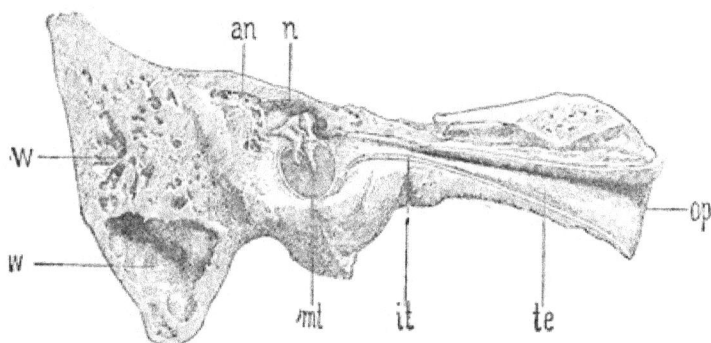

FIG. 61.—Sagittal section through the entire middle ear, outer half. (Left ear.)—op = ost. pharyng. tubæ ; te = canalis tubæ Eust. ; it = isthmus tubæ ; mt = membr. tymp. with the malleus and incus ; n = niche of the body of the malleus and incus ; an = antrum mastoid. ; w w = cells of the mastoid process. After a preparation in my collection.

tympanic membrane, and the malleus, incus, and chorda tympani from the pyramid, with the stapes and the other parts of the ear. is effected, so that the membrana tympani on the one side, and the tympanic cavity on the other, may now be more minutely examined.'

When making sagittal sections through the entire middle ear on a preparation which has been removed from the cranial cavity after the method described on page 8, the osseous Eustachian tube is opened from above, after taking off the tegmen tymp.; while the parts of the sphenoid bone that cover the cartilaginous tube are so far removed with chisel and forceps as to allow the tubal canal, from its pharyngeal orifice to the isthmus, to be split in its entire length.

It will now depend whether, in dividing the temporal bone by a sagittal cut into an outer and an inner half, the cartilaginous Eustachian tube is also to be divided into halves, or whether the

greater portion of the tubal canal is to remain in connection with the
outer or the inner piece of the preparation. In the former case, after
the insertion of a probe into the canal, the roof of the tube (point of
curvature of the cartilaginous hook) is cut longitudinally, with a pair
of straight fine scissors, from the ostium pharyng. to the osseous tube,
the membranous floor of the canal being likewise divided in the same
direction up to the osseous portion, whereupon the cut through the
osseous tube, the tympanic cavity, and the mastoid process is continued with the fret-saw in exactly the same manner as previously
described.

As, however, by this mode of representation the relationship of
the tubal canal to the tympanic cavity fails to be sufficiently
characterized, it is advisable to make also preparations in which
the Eustachian tube is left in connection, some with the outer half

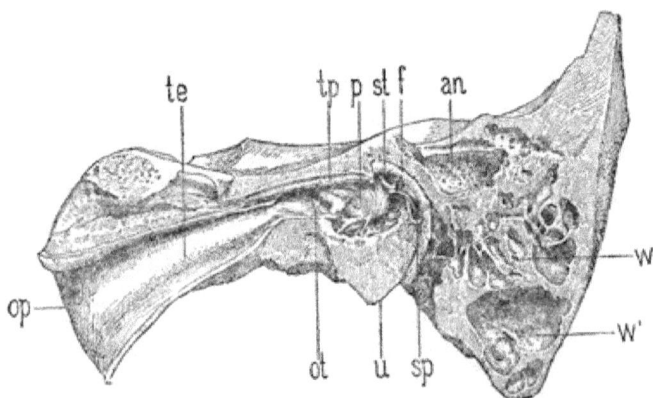

Fig. 62.—Sagittal section through the entire middle ear of an adult, inner half. (Left ear.)
—op = ost. pharyng. tubæ ; te = canalis tubæ Eust. ; ot = ost. tymp. tubæ ; tp = musc.
tens. tymp. ; p = promontory with the anastomos. Jacobsonii ; u = lower wall of the
tympanic cavity ; st = stapes ; sp = musc. stapedius ; f = facial nerve ; an = antrum
mastoid. ; w w' = mastoid cells. After a preparation in my collection.

of the divided temporal bone (Fig. 61), and others with the pyramid
(Fig. 62).

In the first case the saw must be applied medially from the fully-
exposed Eustachian tube, between it and the pyramid, care being
taken, while advancing towards the tympanum, to keep as near as
possible to the inner tympanic wall. In the second case (Fig. 62) the
cut is directed laterally from the Eustachian tube, laid open at the
outer membranous wall, in such a way that it falls in the immediate
vicinity of the external tympanic wall and of the insertion of the
membrana tympani.

The sagittal sections through the middle ear here described are especially suitable for measuring the length dimensions of the Eustachian tube, as well as the distance from the ost. pharyng. tube to the tympanic membrane, to the promontory, and to the entrance into the antr. mastoid. Although the average measurements of the length of the entire Eustachian tube noted by me agree, in the majority of preparations, with those given by Von Tröltsch (35 mm.) and Bezold (36·4 mm.), there are not infrequently cases where the length of tube measures only 33 mm., and, on the other hand, cases where it amounts to 39 mm. Bezold in one case found it to be even 40 mm. In the new-born infant the Eustachian tube is much shorter than in the adult (Von Tröltsch). According to Eitelberg ('Zeitschrift für Ohrenheilkunde,' Bd. 13), its length amounts on an average to 19 mm., of which 11 mm. go to the cartilaginous, and 8 mm. to the osseous portion. The distance from the ost. pharyng. tub. to the anterior edge of the tympanic membrane amounts in the adult, according to my measurements, on an average to 37 mm.; to the promontory, 40 mm.; to the aditus ad antrum, 47-49 mm. An acquaintance with these distances is in a practical respect so far of importance, as when employing bougies for dilating the tubal canal, or when introducing thin elastic tubes (Paukenröhrchen) into the tympanum for the purpose of washing it out, we are enabled to judge, from the length of the inserted portion of the instrument, whereabouts in the middle ear its point may be.

5. PREPARATION OF THE TYMPANIC MEMBRANE.

The anatomical details of the tympanic membrane when it has been separated, together with the malleus and incus, from the pyramid, may be examined either on preparations in which the membrana remains in connection with its osseous frame, or on such where it has been loosened from its insertion in the sulcus tymp. and at the Rivinian segment.

To represent the tympanic membrane, with its osseous frame, as an anatomical preparation, the mastoid portion of the outer half of the divided temporal bone is so fixed in the vice that, during the operation of sawing, the external and internal surfaces of the tympanic membrane can be kept in view.

The external surface of the membrane is first exposed by sawing through, with the finest fret-saw, the inner extremity of the osseous meatus from the front backwards, ½-1 mm. laterally from the insertion of the membrane, and as far as possible parallel to it, the cut being continued 2-3 mm. into the pars mastoid. Next, a cut is carried with the saw, 1½-2 mm. from the insertion of the membrana, round the tympanic frame, commencing at the anterior superior segment, 2 mm. in front of the head of the malleus, and continuing the cut along the anterior, inferior, posterior, and superior periphery of the membrane. The posterior cut passes 2 mm. behind the short process of the incus; the superior, 2 mm. above the malleo-incudal articulation. The preparations thus obtained, and the appearance of which is shown by the accompanying illustrations (Figs. 63 and 64) of the outer and inner surfaces of the tympanic membrane, enlarged to several times its size, are particularly adapted for instruction purposes; but for a thorough anatomical study of the details of the membrane and of the malleo-incudal articulation, as well as of its ligamentous connections with the

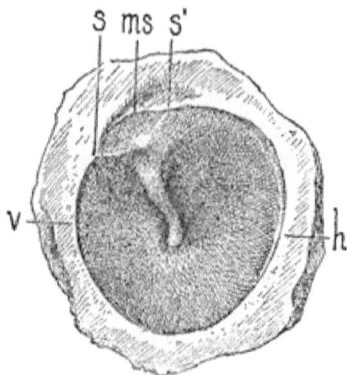

FIG. 63.—Outer surface of the left tympanic membrane of an adult, enlarged 3½ times.—v = segment of the tympanic membrane lying in front of the handle of the malleus; h = posterior segment of the tympanic membrane; s s' = Prussak's striæ, passing from the short process of the malleus to the spina tymp. post. et minor; ms = membrana Shrapnelli.

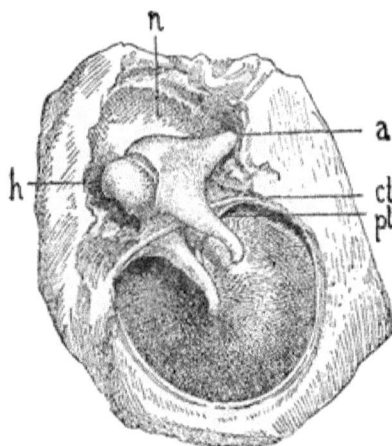

FIG. 64.—Inner surface of the right tympanic membrane with the malleus and incus, enlarged 3½ times.—n = malleo-incudal niche at the outer wall of the tympanic cavity; h = head of the malleus; a = incus; pl = posterior fold of the tympanic membrane with Von Tröltsch's pouch; ct = chorda tympani.

tympanic cavity, other methods of preparation have to be employed, which shall here be discussed.

By laying bare the inner surface of the tympanic membrane, certain pathological changes in it can be more conveniently inspected than is possible by examining its external surface. Among these are the extent and density of calcareous deposits, thickening of the mucous membrane layer, hypertrophy of its fibrous trabecular structure, papillary proliferations; further, the degree of inward curvature of those parts of it which are thinned by cicatrices and atrophy, adhesion of the membrane, or of any cicatrices on it, with the long crus of the incus ; finally, the pathological changes in Von Tröltsch's pouches (accumulation of mucus, connective-tissue adhesions). Thickening of the capsular ligament of the malleo-incudal joint, sometimes occurring in chronic middle ear catarrhs ; again, adhesions and coalescence between malleus and incus with the upper and external tympanic walls, caries and necrosis of the malleus and incus, etc., can be best demonstrated on preparations of this kind.

(a) Loosening of the Membrana Tympani from the Sulcus Tympanicus.

The form and size of the tympanic membrane can only be estimated after it has been prepared out of the sulcus tymp. and incisura Rivini.* Removing the tympanic membrane intact from its frame must, however, if disfiguring rents are to be avoided, be carried out with some care. The simplest way to proceed is as follows : At the inner side of the preparation, close to the anterior margin of the membrana tympani, an incision 1½-2 mm. long is made with a small knife in the mucous membrane. Through this incision the point of the knife is inserted between the edge of the tympanic membrane and the sulcus tymp., and with a few strokes the connection between this and the annulus tendinosus is completely severed. If now the piece of the latter thus loosened be lifted out with the point of the knife from the sulcus tymp., the other portions of the tympanic membrane still remaining in the sulcus can be readily raised. Having firmly grasped with a strong but finely-pointed pincette the free piece of the annulus tendinosus, a gentle pull, aided by the point of the knife, will be sufficient to lift it out complete. This proceeding should, however, only be carried as far as the vicinity of the incisura Rivini, where the sulcus terminates, because further attempts to loosen, by traction, the edge of the tympanic membrane at other points would lead to fibrillation of its upper portion. When the tendinous ring is loosened in front and behind as far as the chorda tymp., the latter is cut through, at the points of its entrance and exit, with a pair of finely-pointed scissors ; the ligaments of the malleus and incus, which will be described later on, being divided at the same time with a small pointed knife.

The tympanic membrane is then, by pressure upon the lower extremity of the manubrium, pushed so far towards the external

---

* Measurements made on the tympanic membrane after it has been prepared out give somewhat higher figures (10-11 mm. for its greatest diameter, 9-10 mm. for its transverse diameter) than those on a preparation where it has been left in its insertion (9½-10 mm. for its greatest diameter, 8½-9 mm. for its transverse diameter). This is explained by the fact that in the latter case the annulus tendinosus lying in the sulcus tymp. is not included in the measurement.

meatus that, after cutting through the ligamentum mallei extern. (see p. 81, Fig. 67), passing from the neck of the malleus to the Rivinian segment, the upper attachment of the tympanic membrane and of the membrana Shrapnelli can be seen. If now, with the point of a very sharp knife, the insertion of Shrapnell's membrane at the Rivinian segment, and the attachment of the superior and lateral portions of the membrane not yet divided, are cut through close to the bone, the connections of the membrana tympani with its surroundings will be completely detached.

Rolled up, and kept connected with the malleus and incus, the membrane unfolds itself rapidly under water, and displays its exact contours and the boundaries of the tendinous ring, which, as a white marginal streak, gradually diminishes in an upward direction, and is entirely lost in the neighbourhood of Shrapnell's membrane. In spite of being detached, the tympanic membrane retains its original funnel-shaped curvature (Helmholtz).

Various sized particles of the lining of the osseous meatus, less frequently of the mucous membrane of the tympanic cavity, becoming detached along with the tympanic membrane, are in all cases found adhering to the preparation. These, which spoil the appearance of the specimen, are best laid hold of under water with a fine pincette, and removed bit by bit, with small curved scissors, close to the tendinous ring.

The tympanic membrane, taken out of its osseous frame, may either serve as an anatomical preparation for demonstration purposes, or be used for microscopical investigation (see the Histological part). In the first case, the membrane is kept in alcohol in small glass jars,* where the details of the outer and inner surfaces of the object can be examined with the aid of a magnifying lens.

The proceeding in disarticulating the pathological tympanic membrane from the sulcus tymp. is the same as with the normal. The separation of the membrane in pathological cases is, however, undertaken only for the purpose of microscopical examination.

If the pathological preparation is to be used as a macroscopic object, it must be left in its natural connection with the sulcus tymp., the morbid changes on the membrane being so better preserved in their original position than would be the case if it were prepared out.

---

* See plate, Fig. IV. belonging to the chapter 'Making Normal and Pathological Preparations for Instruction Purposes.'

(b) *Exposure of the Duplicatures on the Inner Surface of the Membrana Tympani.*

Besides the previously-mentioned details on the inner surface of the tympanic membrane, there are two duplicatures (Fig. 65), one of which, lying behind the handle of the malleus, claims our special attention. It appears as a fold, concave downwards, projects from the membrana, and, partially hidden by the body and long crus of the incus, it passes in a strong curve from the posterior superior periphery of the tympanic frame to the handle of the malleus, and is attached below the middle of the latter. This, the posterior duplicature of the tympanic membrane, which was first described in detail by Von Tröltsch, forms, with the inner surface of the membrane facing it, the so-called posterior tympanic pouch.

To become acquainted with the connection of this duplicature with the upper portion of the tympanic membrane, the incus must be luxated and removed. As, however, by tearing away the incus by force, the delicate ligaments and folds which connect the malleus and incus with the superior and external tympanic walls may be torn, and the preparation rendered useless for the examination of the ligamentous apparatus of the malleus, it is advisable first to cut the ligamentum incudis post. (see p. 82) and the inner side of the capsular ligament of the malleo-incudal articulation, with a small pointed knife, after which the incus can easily be loosened with the pincette.

When the incus has been removed, the posterior duplicature of the membrane becomes visible in its entire extent (Fig. 65). As regards its relationship to the tympanic membrane, it varies greatly in different individuals, and even in one and the same individual differences in both ears are met with. The posterior lower section of the duplicature (c) is not in direct connection with the tympanic membrane, but arises from a slightly curved ledge of bone lying within the sulcus.

Only from 2 to 3 mm. above the lower extremity does the fold pass from the osseous ledge on to the tympanic membrane. Now, while the upper border of the duplicature is frequently found united with the upper periphery of the membrana, its archlike point of attachment appears in other preparations from 1 to 2 mm. lower.

The anterior segment of the duplicature is sometimes divided into two laminæ, the superior of which, connected with the chorda tymp., is attached close to the inner edge of the handle of the malleus, while the inferior lamina is inserted in the angle between the posterior surface of the manubrium and the tympanic membrane—sometimes at a distance of 1 to 1½ mm. from the handle. The triangular space thus formed is generally closed in above, and is not in communication with the tympanic pouch proper.

The posterior pouch communicates, in the preparations which have been examined by me frequently, though not always, with the space above the short process of the malleus (Prussak), which will be described later on ; often, however, its cupola terminates in a funnel-shaped depression. As a rule, Von Tröltsch's pouch is free, yet in some cases delicate bridges or star-shaped ramified bands of connective tissue are found stretched in it, similar to those one meets with in the niches of the oval and round windows, and in the antrum mastoid. Coalescence of the posterior duplicature of the tympanic membrane with the long crus of the incus, in which the chorda tymp. is also

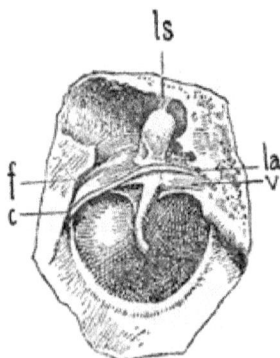

FIG. 65.—Inner surface of the tympanic membrane after removal of the incus.—ls = ligament. mall. super. ; la = ligament. mall. ant. ; f = posterior duplicature of Von Tröltsch ; v = anterior duplicature of the tympanic membrane ; c = chorda tymp. After a preparation in my collection.

frequently involved, as well as adhesions of the surface of the duplicature to that of the membrana, are by no means rare.

The anterior tympanic pouch (Fig. 65 v), smaller in extent, is formed by the spina tymp. posterior, facing the neck of the malleus, and by that fold of mucous membrane which, inferiorly, encloses the ligament. mallei ant., stretched in the direction of the Glaserian fissure, the chorda tymp., and the arteria tymp. inf. The apex of its cupola is almost invariably closed above.

(c) Preparation of the Membrana Shrapnelli.

The external surface of Shrapnell's membrane (Fig. 63) requires no special preparation, since it comes clearly and fully into view when the outer surface of the tympanic membrane is exposed in the regular way. Its rounded edge corresponds to the Rivinian segment (margo tymp., Henle), its inferior boundary being, according to Prussak, formed by two short tightly-stretched striæ (Befestigungs-strang des Hammers, Helmholtz—s s'), which extend from the corners of the incisura Rivini to the tip of the short process of the malleus.

Especially clearly marked are these striæ of Shrapnell's membrane (Fig. 63 s s') in the living subject, while in anatomical preparations they rarely appear so distinct. Schwalbe's opinion (l.c., p. 442), that the posterior one owed its origin to the cutaneous fasciculus of fibres, which passes from the upper wall of the meatus on to the tympanic membrane, I am, from my own investigations, unable to share, since this stria lies deeper in the membrane than the superficial one of the cutis, and can, after removal of the latter, sometimes be demonstrated as a fibrous filament connected with the osseous border of the margo tymp.

In order to gain access to the inner surface of Shrapnell's mem-

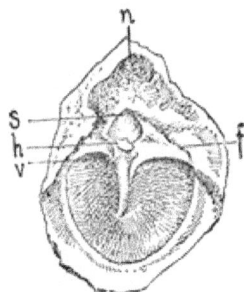

Fig. 66.—Inner surface of the membrana Shrapnelli.—s = membrana Shrapnelli ; f = posterior duplicature of the tympanic membrane ; v = anterior duplicature ; h = section of the neck of the malleus which has been broken away ; n = malleo-incudal niche.

brane (Fig. 66) for the purpose of examination, the incus is first removed, the neck of the malleus being then forced away just above its transition into the upper surface of the short process of the malleus by means of fine forceps.

If now the remaining portion of the ligament. mallei extern., which passes from the neck of the malleus to the outer tympanic wall, and forms the superior wall of Prussak's space, be carefully removed with a fine pincette, the inner surface of Shrapnell's membrane is laid bare, and with the aid of a lens its boundary may be determined on all sides.

The arch-like edge of Shrapnell's membrane is partly in immediate connection with the osseous border of the incisura Rivini, and here serves as the periosteum. Destruction of the membrane up to its extreme edge, therefore, frequently leads to carious fusion of the osseous wall adjoining the Rivinian segment, and to the formation of

large gaps in the upper and posterior walls of the meatus, thus allowing a clear view into the upper tympanic space.

(d) *Representation of the Ligamentous Apparatus in the Vicinity
of the Malleus and Incus, and of the System of Cavities between
the Body of the Malleus and Incus, and the External Wall of
the Tympanic Cavity.*

Of late years special attention has been directed to the anatomical
relations of that space which lies between the external surface of the
body of the malleus and incus and the smooth niche of the external
tympanic wall facing it. Their great practical importance is due to
the fact that purulent processes frequently set up in this space, which
are accompanied by perforation of Shrapnell's membrane, and are
characterized both by their obstinate course and frequent complica-
tions with suppurations in the mastoid process. An exact acquaintance with this part of the middle ear is, therefore, of special importance to the aurist, and we here propose to enter more fully into the
discussion of the method of representing the anatomical details of this
space.

The ligamentous apparatus between the malleus and incus and the
adjacent tympanic walls consists partly of constant, compact, fibrous
ligaments, partly of inconstant, delicate folds of mucous membrane,
connective-tissue bands, and delicate threads.

The laying bare of the constant fibrous ligaments of the malleus
and incus presents no difficulty, since they appear, on fresh as well
as on spirit preparations, strongly marked by their whitish tendinous
appearance.

1. The ligamentum mallei superius (Fig. 69, l s), a roundish or flat
band passing from the upper tympanic wall to the head of the malleus,
is exposed in such a manner that, after removal of the tegmen mastoid.,
the roof of the tympanic cavity is taken away with the pointed bone
forceps from behind forwards, care being taken to avoid breaking off
that portion of the tegmen tymp. which lies above the body of the
malleus and incus. This is the more safely accomplished if, on the
body of the incus becoming visible, we keep medially from it. When
the tympanum has been opened in front so far as to allow the tensor
tendon (Fig. 59, s) to be plainly seen, it will be easy, by holding the
preparation slightly to one side, to find the ligament. mall. sup. If the
pyramid be now also separated from the preparation, both the origin
and insertion of this ligament can be clearly seen. Sometimes it arises
from a projection of bone on the upper tympanic wall bulging forward
towards the head of the malleus. Almost constantly, there are

attached to the ligamentum mallei superius, in front, behind, and laterally, transparent folds of mucous membrane, which will be considered more in detail in a future chapter.

This ligament is a check ligament for the excursions of the handle of the malleus outwards. Not infrequently there is, in the normal state, direct contact between the head of the malleus and the tegmen tymp., in which case the ligamentum mallei superius cannot be demonstrated. This anatomical relation favours, in cases of inflammation in the upper tympanic space, adhesions of the head of the malleus with the upper tympanic wall.

2. The ligamentum mallei anterius (Fig. 67), a strong, tendinous band, which arises partly from the spina angularis of the sphenoid bone (Henle), partly from the spina tympanica posterior (spina tym-

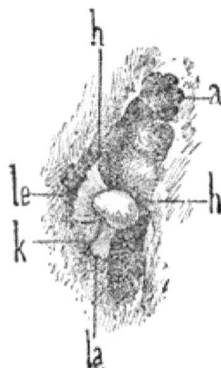

Fig. 67.—Ligament. mallei anter. et extern. (Right ear.)—h = head of the malleus ; la = ligament. mallei anter. ; le = ligament. mallei extern. ; h = its posterior portion ; k = osseous tip of the spina tympan. post. (major), projecting between the ligament. mall. ant. et extern. ; a = antrum mast. After a preparation in my collection.

panica major, Helmholtz), and the border of the Glaserian fissure ; it surrounds in the new-born infant the processus Folianus of the malleus, but embraces in the adult merely the stump of this process, and finds its insertion at the neck of the malleus and the lower portion of the anterior and lateral surfaces of the head of the malleus. This short, tense, though somewhat broad ligament becomes visible to its full extent after removal of the chorda tymp., and of the fold of mucous membrane which is stretched between the anterior surface of the head of the malleus and the adjoining tympanic wall. That portion of the ligament which is situated in the Glaserian fissure, cannot be seen until after removal of the piece of the external wall of the tympanum, and of the Eustachian tube lying in front of the sulcus tymp. and the fissura Glaseri, and is easily represented on decalcified preparations and microscopical sections.

The important influence of the ligament. mall. anter. upon the position of the manubrium, and consequently upon the tension of the tympanic membrane, may be seen from the following experiments made by me. On cutting the tendon of the tensor tympani and aspirating the air from the external meatus, the tympanic membrane will be drawn outwards, but return to its former position as soon as the aspiration ceases. If, now, the incudo-stapedial articulation be severed, the incus detached from the malleus, and the ligament. mallei super. et ext. cut through, the manubrium will move distinctly in an outward direction, but, pressed towards the external meatus, it will still spring strongly inwards. Not until the ligament. mallei anter. is cut through, does this springing of the manubrium inwards entirely cease.

3. The ligamentum mallei externum, a tendinous fan-shaped ligament (Fig. 67, le), connects the upper posterior portion of the incisura Rivini with the crista capitis mallei opposite ; it is, therefore, a check ligament for the movements of the handle of the malleus outwards. The posterior part of this ligament (h) has been termed by Helmholtz ligamentum mallei posticum, and as its line of direction, prolonged anteriorly by the malleus, meets the middle fibrous filaments of the ligamentum mallei anterius, he calls these two fibrous fasciculi the axis ligament of the malleus. To demonstrate this ligament the tegmen tymp. is completely removed up to its lateral boundary, the incus is also taken away, and the inconstant folds possibly remaining between the head of the malleus and the external tympanic wall are removed with the pincette. The removal of the incus allows, in the majority of preparations, the boundaries of the tendinous, gray and sharply-defined ligament to be seen. Where, owing to the narrowness of the space between the malleus and outer tympanic wall, this is not possible, the ligament is best exposed by breaking off the head of the malleus above the neck with a pair of fine forceps.

According to Schwalbe (l.c., p. 504) the ligament. mall. ant. is separated from the ligament. mall. ext. by a bridge of mucous membrane, which, with the latter, forms the upper boundary of Prussak's space.

In several cases I found this interspace occupied by the inwardly projecting spina tymp. post. (compare Fig. 67, k).

4. The ligamentum incudis posterius (Fig. 68), which connects the short process of the incus, covered with a layer of cartilage, with the corresponding saddle-shaped and similarly covered niche on the posterior tympanic wall, may be recognised without further preparation, with the naked eye, or by the aid of a magnifying lens, as a whitish tendinous band surrounding, in the form of a fan, the short process of the incus.

Not infrequently the middle portion of the ligament corresponding to the tip of the short process is wanting, in which case (Fig. 68) the latter is found to be fixed by a lateral (b') and a medial (b) ligament.*

Besides the ligaments here described, the malleus and incus are connected with the walls of the tympanic cavity by a number of partly constant, partly inconstant folds of mucous membrane. The most constant of these is the lateral incudal fold (Fig. 68, f), which passes from the upper edge of the body of the incus or its external surface to the lateral wall of the niche, and after removal of the tegmen tymp. becomes clearly visible. It is connected behind with the ligam. incud. ext. (b'), in front, though not constantly, with the fold (lateral malleo-incudal fold), passing from the head of the malleus to the wall of the niche. In the latter case the space between the body of the malleus and incus and the lateral niche is found to be

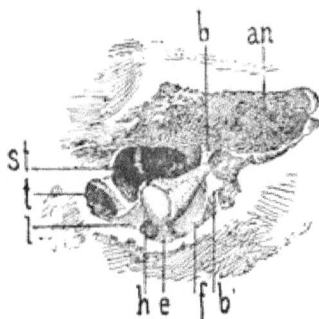

FIG. 68.—Ligamentous apparatus of the malleus and incus.—h = head of the malleus ; l = ligament. mall. ant. ; e = ligament. mall. ext. ; f = outer fold of the incus ; b = inner, b' = outer portion of the ligament. incud. post. ; t = tendon of the musc. tens. tymp. ; st = incudo-stapedial connection ; an = antr. mast.   After a preparation in my collection.

completely closed above. Frequently, however, there exists, between the incudal fold and the lateral fold of the malleus, a gap, varying in size, and in the incudal fold itself one or several openings are sometimes met with.

Among the constant bridges of mucous membrane in the vicinity of the malleus and incus, there are further to be numbered : one, passing from the crista transv. tymp. (p. 47) to the tendon of the tensor tympani, and one stretched horizontally (Zaufal) from the latter to the anterior mucous membrane duplicature of the tympanic membrane. Not constantly, though frequently enough, broad bridges of mucous membrane or delicate filaments are found between the long process of the incus and the inner tympanic wall ; these are the

* Compare Sömmering, Abbildungen des menschlichen Gehörorgans, Taf. 2, Fig. 20.

residue of the mucous membrane fold, which at this spot occurs in the embryo constantly (Urbantschitsch).   We further notice : mucous-membrane bridges between the manubrium and the long crus of the incus; a fold extended between the short and the long process of the incus; then, bridges and filaments between the malleus, the long crus of the incus, and the stapes ; and, lastly, a ramified trabecular structure stretched from the body of the malleus and incus into the antr. mastoid., with the importance of which structure we shall become acquainted hereafter.   That these folds of mucous membrane, frequently occurring in great numbers in the tympanic cavity, favour in chronic middle ear catarrhs the setting up of adhesive processes has already been mentioned.

The space situated between the body of the malleus and incus and the lateral wall of the niche, and bounded above by the malleo-incudal fold, below by the ligament. mall. ext., is also crossed by inconstant folds of mucous membrane, which are more numerous in the new-born infant than in the adult.   The fold most frequently occurring here is the vertical fold of the niche (Fig. 70 w s), which, stretched between the malleus or malleo-incudal articulation and the niche, is in connection below with the ligament. mall. ext., above with the lateral malleo-incudal fold.   It becomes visible after removal of the incus, or after breaking away the head of the malleus (in glycerine preparations).   By this vertical fold a space is bounded off laterally, in front of the head of the malleus, and above Prussak's space (see later on), which communicates, through one or several apertures, with the cavum tymp. and the space of Prussak.   It is in this space especially that those obstinate purulent processes, which accompany perforation of Shrapnell's membrane, establish themselves, and which often resist the most energetic treatment.

(e) Representation of Prussak's Space, situated above the Short Process of the Malleus.

The space above the short process of the malleus (Fig. 69), first described by Prussak, is formed externally by Shrapnell's membrane (Fig. 63, p. 74), below, by the upper surface of the short process of the malleus, as well as by the cupola of the posterior and anterior tympanic pouches, and above by the ligament. mall. ext.   The exact limits of this cavity can, however, only be shown on sections of decalcified preparations, or in a series of microscopic sections.   For the purpose of macroscopic representation, the tympanic membrane and its osseous frame, detached along with the malleus and incus from the pyramid, must be decalcified in a mixture of common salt and nitric acid

(5-10 %), then well washed in water, and hardened in alcohol. In order to avoid the displacement of the various parts of the preparation during the cutting, the depression formed by the external surface of the tympanic membrane, and the adjoining remains of the osseous meatus, is filled with melted paraffin, and after the setting of the latter it is cut away on the flat, so that in turning the preparation over, the inner surface of the tympanic membrane should come to lie upwards, and in a horizontal position. With a sharp razor used for microscopic purposes, a cut is now carried through the whole of the preparation in the longitudinal direction of the malleus, exactly in the middle line of its head and handle. After this the paraffin is carefully removed, and the sections of the preparation are examined under water or in alcohol.

If the cut has been carried exactly through the short process of the malleus, we notice (Fig. 69) on the anterior surface of the section the boundaries of Prussak's space (o), sharply defined above by the ligament. mall. ext. (le); internally, by the neck of the malleus; below, by the upper surface of the short process of the malleus (s); and externally, by Shrapnell's membrane (o s). The anterior part of this space is mostly closed on all sides, but sometimes communicates, between the body of the malleus and incus and the lateral wall of the niche, with the space (Fig. 69, r) which is situated above the ligament. mall. ext. The posterior part communicates frequently, though not always, by a small aperture, with the posterior tympanic pouch; but one often finds also, on the posterior membranous wall of this space, one or several apertures, which, after removal of the incus, can be distinctly seen by holding the preparation a little to one side, and through which there is communication between Prussak's cavity and the space situated between the body of the incus and external tympanic wall, and freely connected below with the tympanic cavity.*

Most suitable for the examination of the ligamentous apparatus of the malleus and incus are preparations which are either quite fresh, or such as have been previously hardened in alcohol and afterwards soaked in glycerine. Wet preparations which have been kept in spirit are, for the examination of the ligaments and folds, quite as unsuitable as dry objects, on which these parts become fissured through shrinking. To represent preparations of this kind which are to be kept for demonstration, I make use of spirit preparations, over which I pour, on taking them out of the liquid, a small quantity of glycerine

* Kretschmann's treatise on 'Fistulous openings at the superior pole of the membrana tympani' (Arch. f. Ohrenheilk, Bd. 25), which appeared while this chapter was in the press, could not be noticed here.

S6 EXAMINATION OF LIGAM. APPARATUS OF THE MALLEUS AND INCUS.

sublimate (0,1:25,0), drying them afterwards in a place kept free
from dust. Treated thus, even the most delicate folds of mucous
membrane retain for years their natural form and position. Such
objects are best protected from dust when fixed on a small stand
provided with a glass shade (Plate Figs. I. and III.). Equally
instructive are sections of decalcified preparations, which, for the
purpose of representing the ligamentous apparatus on transverse
sections, should be made in a vertical as well as in a horizontal
direction to the malleo-incudal articulation. As, however, the colour
of the ligaments and folds on decalcified spirit preparations differs
but little from that of the surrounding bone, it is advisable to expose
fresh objects for several weeks to the action of Müller's fluid (see
Histological part), or of a weak solution of chromic acid, and only
after such treatment to proceed with the decalcification by means of

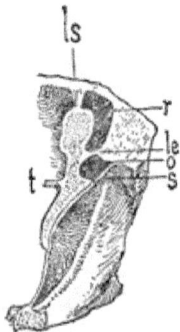

FIG. 69.—The space of Prussak: section
through the tympanic membrane, malleus,
upper and outer tympanic wall of a decalcified preparation. — ls = ligament. mall.
super. ; le = ligament. mall. ext. ; s =
membrana Shrapnelli ; o = Prussak's
space ; r = system of cavities between the
body of the malleus and incus, and the
external tympanic wall ; t = tendon of the
musc. tens. tymp. After a preparation
in my collection.

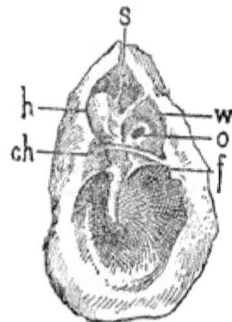

FIG. 70.—Elliptic opening of Prussak's space
on its posterior membranous wall after removal of the incus.—h = head of the
malleus ; s = ligament. mall. super. ;
w s = Vertical fold of mucous membrane
between the malleus and the malleo-incudal niche ; o = elliptic opening of
Prussak's space ; ch = chorda tympani ;
f = posterior duplicature of the tympanic
membrane. After a preparation in my
collection.

hydrochloric or nitric acid. In sections of such preparations the
osseous tissue appears dark green, while ligaments and folds are
distinguished by their whitish or light-green colour. To avoid tearing
the delicate folds of mucous membrane when making sections, the
decalcified preparation, after having previously been freed from
water in 90 % alcohol, is soaked in celloidin solution ; it is
then placed in dilute alcohol, and after setting of the medium the
sections are made (see 'Celloidin Embedding,' in the Histological part).

By the ligamentous apparatus of the malleus and incus here

sketched, and the folds of mucous membrane and bridges stretched between them and the upper external tympanic wall, there is formed in the outer portion of the upper tympanic space, between the external surface of the malleus and incus and the niche opposite, a system of cavities, first described by me (Wien. med. Wochenschrift, 1868), the spaces of which communicate with each other and with the rest of the tympanum. In new-born infants I found, on microscopical sections of this system of cavities, in some of its loculaments, a yellow serous or slimy fluid; in the adult all the spaces of the meshwork contain air.

The pathological changes which are developed in this region of the upper tympanic space have hitherto not been sufficiently investigated —at least, not to such an extent as the importance of the subject would demand. It must, however, be taken as an undisputed fact that the spaces of this system of cavities in chronic mucous middle ear catarrhs, running on without perforation of the tympanic membrane, are filled with succulent neoplastic connective tissue, and that similar changes are sometimes met with at this spot as the result of purulent inflammation. The latter, however, not infrequently leads to destructive changes, especially to perforation of Shrapnell's membrane, fusion of the margo tymp., and to carious and necrotic destruction of the malleus and incus.

The purulent process in the above-mentioned system of cavities accompanied by perforation of Shrapnell's membrane, may either localize itself as an independent process without spreading to the rest of the tympanic cavity, or—and this is more frequently the case—the entire tympanic cavity is attacked from the commencement by the suppurative inflammation. In the course of the process, however, the suppuration in the middle ear may cease, while it obstinately continues in the system of cavities between the body of the malleus and incus and Shrapnell's membrane. It admits of explanation, that in a considerable number of cases pus may be copiously secreted in these cavities, and make its way through the perforated Shrapnell's membrane, without it being possible to prove the presence of any secretion in the rest of the tympanic space, when it is considered that the scanty communications of this system of cavities with the tympanum are obstructed through swelling, inspissated secretion, and cholesteatomatous masses, by proliferations of mucous membrane, and by polypi (Morpurgo). We can also conceive the obstinacy of a purulent process in a space occupied by a meshwork from which the stagnating secretion is difficult to remove, and the hidden crevices of which are only with difficulty accessible to antiseptic treatment.

As regards the connection of purulent processes in this space with simultaneous chronic suppuration in the antrum mastoid., there are, in my opinion, two things of importance to be considered. We have seen (p. 38) that the smooth surface of the malleo-incudal niche merges behind into the outer cellular wall of the antrum mastoid. Purulent inflammations in that niche may thus, with equal facility, spread *ex contiguo* to the neighbouring mucous membrane of the antrum mastoid.; and, on the other hand, those occurring in the antrum may find their way to the malleo-incudal niche. The latter might, according to clinical observations, be much more frequently the case than has hitherto been supposed; thus purulent discharges, with perforation of Shrapnell's membrane, would in such cases have to be looked upon as the consequence of suppuration in the antrum mastoid.

Another circumstance respecting the connection of purulent processes in the antrum mastoid. with perforation of Shrapnell's membrane, which so far has not been sufficiently appreciated, lies, in my opinion, in the peculiarity of the inclination of the external wall of the antrum and of the malleo-incudal niche. The strong inclination of these osseous walls to the horizontal may be easily demonstrated on macerated as well as on unmacerated skulls, after removal of the tegmen tymp. The outer tympanic wall, with the malleo-incudal niche and the external wall of the antrum, must therefore, in consequence of their strong inclination, be designated the outer inferior wall of the middle ear; and it is thus evident that the purulent secretion, flowing from the antrum at its outer inferior wall, must first reach the malleo-incudal niche. This will happen especially in those cases where, in the external malleo-incudal fold, either normal interstices are found, or where breaches have been caused in it by the purulent process. In such cases the pus, flowing out of the antrum, must reach the system of cavities between the body of the malleus and incus and the external tympanic wall, setting up suppurative inflammation therein, and eventually leading to perforation of Shrapnell's membrane. This spreading of the suppuration from the antrum to the system of cavities will be favoured if, at the same time, the communication between the antrum and the tympanic cavity proper be interrupted by normal or pathological bridges of connective tissue.

It has further been pointed out that in preparations of the normal ear there is sometimes stretched out in the antrum mastoid. a trabecular structure connected with the osseous walls, and composed of ramified bands of mucous membrane, which, as seen in several preparations of my collection, extends into the tympanic cavity, and is there connected with the ligaments and mucous membrane folds of the malleus and incus. Now, it may be assumed, with some probability, that where such normal formations occur in the middle ear, the changes caused by inflammation in the vicinity of the malleo-incudal niche may, through this trabecular structure, spread into the antrum, and, in an opposite direction, from the antrum into the niche. This was confirmed by a post-mortem case which came under my notice. An individual who during life had shown symptoms of a chronic mucous middle ear catarrh, was found to have died in consequence of marasmus senilis. Examination of the left ear showed a viscid, yellowish, transparent, elongated plug of mucus which reached from the malleo-incudal niche into the antrum mast., and which, upon

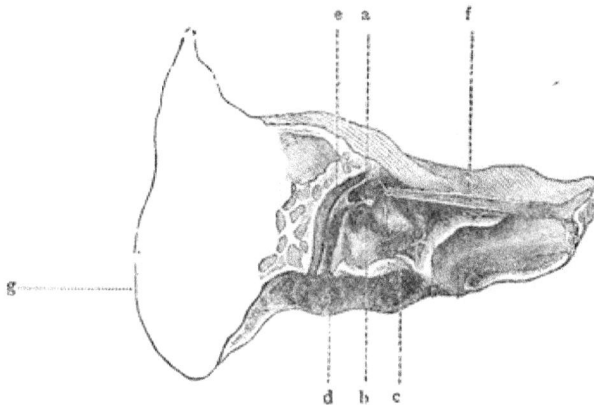

FIG. 71.—Inner wall of the tympanic cavity.—a = fenestra ovalis with the stapes; b = fenestra rotunda; c = promontory; d = musc. stapedius; e = canalis Fallopii; f = canalis pro tensore tympani; g = mastoid process.

closer examination, proved to be the above trabecular structure, the spaces of which were occupied throughout by the viscid mucous mass.

6. INNER WALL OF THE TYMPANIC CAVITY : ITS EXAMINATION IN PATHOLOGICAL DISSECTIONS.

The pyramid, when separated from the pars tymp. et squamosa, presents on its outer anterior wall, turned towards the tympanic cavity (Fig. 71), the two physiologically important labyrinthine fenestræ The fenestra ovalis (a), leading into the vestibular space of the labyrinth, lies, with its strongly inclined plane, at the bottom of a deep niche (pelvis ovalis), and serves for the reception of the stapes. The long diameter of the fenestra ovalis measures rather more than 3 mm., the width being 1½ mm. Below the niche of this fenestra a second, nearly triangular niche, which is considerably inclined backwards, leads to the membrane of the fenestra rotunda, the latter closing the lower scala of the cochlea (scala tymp.) towards the tympanic cavity. In front of the two labyrinthine fenestræ lies the

most prominent part of the inner tympanic wall, corresponding to the
projection of the first cochlear convolution—the promontory. Over
this prominence passes, in a vertical direction, in an open or covered
groove, Jacobson's nerve, which effects the anastomosis between the
ganglion jugulare and the nervus petrosus superficialis minor.

Around the promontory and the two labyrinthine fenestræ are
grouped : superiorly, above the fenestra ovalis, in a direction slightly
backwards and downwards, the middle portion of the Fallopian canal,
with the facial nerve; near this, in a backward direction, and projecting towards the tympanic cavity and the antrum mast., the
external crus of the horizontal semicircular canal; behind the promontory, the eminentia pyramidalis enclosing the musc. stapedius,
and bounded medially and posteriorly by the descending portion of
the facial nerve. In front, the inner tympanic wall, becoming
gradually flatter from the promontory towards the ostium tymp. tubæ,
appears narrowed to a triangular space between the canalis pro tens.
tymp. (f) and the obliquely ascending anterior tympanic wall.

In pathological dissections, after the pyramid has been separated,
attention should first be directed to the condition of the mucous
membrane of the promontory and of the contiguous parts of the inner
tympanic wall. In acute inflammations, as well as in chronic suppurations of the middle ear, the nature of the exudation lying upon the
surface of the mucous membrane, and in the depressions of the inner
tympanic wall, or of any epithelial deposits, should first be examined,
and after these have been carefully washed away, the degree of
hyperæmia, of swelling, and proliferation of the mucous membrane
in the various parts of the tympanic cavity must also be ascertained.

Further, it should be ascertained by means of a lens whether the
mucous membrane, in spite of considerable swelling, presents a smooth
surface, or whether it be covered with granulations and papillary
proliferations. Partial thickening of the mucous membrane, circumscribed infiltrations and loss of substance, may also be demonstrated
by magnifying power. In addition to this, wherever, in purulent
middle ear processes, proliferations of the mucous membrane are met
with on the inner tympanic wall, or where, during life, appearances
of a carious affection of the temporal bone had existed, it must be
determined by careful sounding whether, and to what extent, roughness of the promontorial wall of the tympanic cavity exists. In
extensive caries of the pyramid—further, in tubercular phthisis of the
lining membrane of the tympanic cavity—the extent of the ulceration
in the membrane and the bone will be perceptible in most cases with
the naked eye.

In non-purulent affections of the middle ear, and after suppura-

tions which have ceased, where the lining membrane of the promontory
appears at times pale yellow, atrophic, at other times tendinous, gray
and thickened, an examination should be made with a pointed preparing needle, whether it be somewhat movable over the bone, as in
the normal state, or whether it be firmly adherent to it.

Of great importance in forming an estimate of the disturbances
of hearing in morbid changes on the inner tympanic wall is the
examination of the niches of the two labyrinthine fenestræ. First, as
regards the niche of the fenestra ovalis, this is often found so free
that it is quite possible to see on its floor the external surface of the
foot-plate of the stapes, together with the points of attachment of
both its crura. In not a few cases, however, one finds in this niche,
even in the normal state, especially by magnifying power, delicate
filaments and fenestrated membranes stretched out in varying number,
by which the crura of the stapes are connected with the walls of the
niche.

In examining a large number of normal ears, great differences are
found to exist as to the capacity of the niche of the fenestra ovalis ;
sometimes it appears so spacious, and the distance of the crura of
the stapes from the walls of the niche so great, that it can be
accurately measured with the naked eye ; at other times, again, it is
strikingly narrowed, and the crura of the stapes are in immediate
contact with the inferior or lateral walls of the niche. There is no
doubt that such a congenital narrowness of the niche of the fenestra
ovalis favours, in inflammatory processes of the lining membrane of
the middle ear, adhesions of the crura of the stapes with the niche
of this fenestra.

The pathological results of this niche in diseases of the middle
ear vary according to the duration and peculiarity of the morbid
process. In recent mucous catarrhs the niche is often found filled
with viscid, glassy mucus, while the lining membrane of the niche,
as well as that of the rest of the tympanic cavity, are but slightly
softened and little injected. In acute purulent inflammations, on the
contrary, after removal of the purulent, or muco-purulent exudation
from the niche, its strongly reddened mucous lining shows considerable
tumefaction. Where the previously described connective-tissue filaments and bridges are present in the niche, in acute inflammations,
viscid purulent exudation is retained in the interspaces of this network, and the stapes in such cases appears as though enveloped in
a yellowish-red mass filling up the niche.

In chronic non-purulent middle ear catarrhs, in the course of
which not infrequently connective-tissue proliferations are developed
in the neighbourhood of the ossicula, especially often in the malleo-

incudal niche and in the upper tympanic space, the pelvis ovalis also is frequently found filled with succulent, grayish-red connective tissue, and the crura of the stapes absolutely embedded it it. They constitute the foundation of pathological cord-like formations by which the crura of the stapes become abnormally fixed to the walls of the niche and the stapes loses its power of oscillation.

The most striking alterations in the niche of the fenestra ovalis are met with in chronic suppurations of the middle ear. Here, in most cases, in consequence of round-cell proliferation in the inflamed lining membrane, the niche is so filled up by the dark-red or grayish-red proliferation, that only the capitulum of the stapes projects above the surface of the obliterated niche. After the cessation of middle - ear suppurations, there not infrequently remain behind in the pelvis ovalis grayish-yellow organized masses of connective tissue, and the position of the almost completely obliterated niche is only perceptible by the prominating capitulum of the stapes, and by the ledge-like projecting tendon of the musc. stapedius behind it. Should the crura of the stapes and the capitulum stapedis be destroyed through the preceding purulent process, the obliterated niche of the fenestra ovalis, the level of which reaches that of the rest of the promontorial wall, can only be determined by its relative position to the fenestra rotunda and to the canalis Fallopii.

In dissections, where during life the diagnosis of anchylosis of the stapes was made, and where this was proved by the experiment described on p. 61, the mobility of the stapes can also be tested with the probe after separation of the tympanic membrane from the pyramid. In the normal condition a slight lateral pressure on the capitulum will be sufficient to prove the mobility of this small bone. In rigidity of the stapes in the fenestra ovalis, the mobility is markedly diminished, in anchylosis entirely destroyed. The examination with the probe must, however, be made with great care, because with even slight pressure the crura of the stapes break off, and the preparation is rendered useless for demonstration purposes and for histological investigation.

Where anchylosis of the stapes has been proved to exist, we have to examine whether it is due to adhesion of the crura with the niche of the fenestra ovalis (anchylosis of the crura of the stapes), or to adhesion of the edges of the foot-plate with the circumference of the oval window (anchylosis of the foot-plate of the stapes). In the former case the adhesion of the crura occurs, as a rule, with the inferior wall of the niche, and this may be easily seen with the naked eye or with a lens. In two cases only of congenital deaf-mutism I found, simultaneous with adhesion of the long process of the incus with the posterior tympanic wall, the crura of the stapes adherent to the superior wall of the pelvis ovalis.

The adhesion of the crura of the stapes with the niche of the oval window is either one of connective tissue or an osseous one. The former has been termed anchylosis spuria stapedis. A distinction between the two forms is only possible if the preparation be either decalcified and submitted to microscopical examination (see, with reference to this, the Histological section), or if it has been completely macerated. In the latter case, in adhesion of the crura by connective tissue with the niche of the oval window, the stapes will fall out; in bony anchylosis, however, it will remain firmly attached to the bone. In bony anchylosis of the foot-plate in the oval window, the stapes in the macerated preparation remains inseparably fixed in its position.

In anchylosis of the foot-plate of the stapes, an accurate examination of it, and of the simultaneous changes in the ligamentum orbiculare stapedis, is only possible if the labyrinthine side of the stapes be exposed. For this purpose a cut with a fret-saw is carried in the long axis of the pyramid, parallel to the inner tympanic wall, through the middle of the petrous bone, thus dividing the vestibule into an outer and an inner half. After washing the preparation, the outlines of the inner surface of the foot-plate of the stapes in the fenestra ovalis can now be plainly recognised on the external wall of the labyrinth. If a normal preparation be held against the light, the contours of the transparent ligamentum orbiculare stapedis will come clearly into view. In calcification of this ligament, and in bony anchylosis with the oval window, on the contrary, it becomes opaque, like the adjoining bone. On the side of the labyrinth sometimes the circumference of the oval window is found covered by a protuberant osseous proliferation (Toynbee), or, as in a case observed by me, the entire fenestra is, on the labyrinthine side, covered by osseous neoplasm.

Equally important in pathological dissections is the accurate examination of the niche of the fenestra rotunda, inasmuch as pathological products there diminish the power of vibration of the membr. fenestr. rotundæ, and may impair hearing to a great extent. Since, as before mentioned, the entrance into the niche is directed markedly backwards, a view into its floor can only be obtained by partially removing, with the forceps, the posterior and inferior tympanic walls facing it. With a favourable light, the externally somewhat concave membrane of the round window can be completely seen as it lies stretched out on the floor of the niche. If the membrane be transparent, the lamina spiralis membranacea behind it shines through as a dark, slightly wavy line. Frequently, however, there are found on the membrane gray, ramified trabeculæ of connective tissue (Bezold), such as are met with on the inner surface of the tympanic membrane and at various spots in the lining membrane of the middle ear. On several of the preparations examined by me I further found the entrance to the niche covered by a band of mucous membrane with one or more perforations, after the removal of which the membrane of the round window came into view. That the niche of this fenestra also shows, in normal preparations, many variations with regard to capacity and inclination of its plane to the axis of the ear, is known.

Respecting the morbid changes in the niche of the fenestra rotunda, we find it in catarrhal processes filled with serous fluid, or with tenacious viscid mucus. In purulent inflammations with tough muco-purulent exudation, very frequently a plug of this material is met with filling up the whole of the niche, the removal of which with the pincette is the more difficult the more firmly the exudation is retained in the niche by bands of connective tissue or folds of mucous membrane. Such plugs in the niche of the round window cause impairment of hearing of a high degree, as I have convinced myself by dissections of cases observed during life.

In chronic non-purulent processes on the lining membrane of the middle ear which are accompanied with new formations of connective tissue, most varied alterations are met with in the niche of the fenestra rotunda. Sometimes its membrane shows simply a thickening of its mucous covering; at other times, again, the niche is hidden and obliterated by succulent, tendinous, gray connective tissue; lastly, as rare results of chronic middle ear processes, terminating in sclerosis of the mucous membrane, we observe calcification and ossification of the membrana fenestræ rotundæ (Toynbee, Von Tröltsch), and narrowing or osseous closure of the niche of this fenestra (Moos, Politzer), resulting from hyperostosis of the bony walls.

Almost constantly pathological changes in the niche of the fenestra rotunda are met with in chronic purulent middle ear processes, in which, as is well known, the mucous membrane of the whole of the middle ear undergoes hypertrophy through round-cell infiltration and connective-tissue neoplasms. Accordingly, the niche of the fenestra rotunda appears also to be filled up either by dark-red proliferation of mucous membrane or granulation tissue, or by a small polypus (case of the author's), or after suppuration which has ceased, covered over and obliterated by yellowish-gray, glistening connective tissue.

7. OSSICULA.

The isolated ossicles are best obtained after maceration of the temporal bone, out of which they drop when slightly shaken. Of the three small bones, the one which is most frequently missed is the stapes, which often, after maceration of its annular ligament, falls into the vestibule, out of which it can only be recovered with difficulty by means of a fine pincette. The long process of the malleus (process. Folianus, Fig. 72, d) is entirely preserved on a malleus obtained through maceration of the temporal bone of the new-born infant, while in the adult merely a short pointed stump remains.

The axis of the head of the malleus is inclined to the handle at an obtuse angle. On the posterior surface of the oblong head of the malleus we notice a longish oval articular surface bounded by a slight osseous projection, which passes from above and without downwards and inwards to the border of the neck. Two surfaces are distinguished

FIG. 72. — Malleus. — a = head ; b = neck ; c = handle ; d = long process ; e = articular surface.

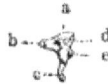

FIG. 73.—Incus.—a=body ; b = short process ; c = long process ; d =articular surface ; e = lower check cog.

FIG. 74. — Stapes. — a = capitulum ; b = crus ; c = foot-plate.

on it, which meet at a nearly vertical edge. Corresponding to this, we find on the body of the incus an articular surface made up of two surfaces, the upper part of which (Fig. 73, d) is directed inwards, while the lower part (e) faces laterally. Helmholtz compares the mechanism of the malleo-incudal articulation with the check contrivance in the interior of a watch-key. With each movement of the manubrium inwards the lower cog of the malleus (Fig. 72, e), will fall into the lower one of the incus (Fig. 73, e), whereby the long process of the incus is made to follow the movement inwards of the manubrium. With the movement outwards of the latter, on the other hand, the lower cog of the malleus will move away from the lower one of the incus, so that the latter will only to a slight degree follow the movement outwards of the malleus. This peculiarity of the malleo-incudal articulation may be proved experimentally by alternately condensing and rarefying the air by means of an india-rubber tube fitting tightly into the external meatus on a fresh preparation from which the tegmen tymp. has been removed, and observing, while doing so, the movements of the manubrium and of the long process of the incus. The action of the lower cog of the head of the malleus may also be demonstrated on macerated ossicles by fixing with sealing-wax, vertically to the long axis of the malleus, a match upon the stump of the long process, and connecting in a like manner also the short process of the incus with another match. If now the articular surfaces of the malleus and incus, by means of these little sticks, be fitted one into the other, and the malleus be alternately turned inwards and outwards, we can convince ourselves of the correctness of Helmholtz's statement above alluded to.

To artificially join the various ossicles into a normal chain, it is best to make use of liquid glue (ichthyocolla), such as is used for mending broken glass, china, bone, etc. One drop is sufficient to join the articular surfaces of the malleus and incus so firmly that it is difficult to separate them afterwards. Less simple is the joining of the capitulum of the stapes to the long process of the incus, on which, in consequence of the maceration, the ossiculum lenticulare Sylvii is mostly wanting. It can only be done successfully if, after bringing the two articular surfaces into contact, the stapes, by means of a fine pincette, be held in its normal position (5-10 minutes) until the cement has dried.

The ossicular chain in its continuity, together with its capsular ligaments, is best represented on preparations which have been hardened in spirits of wine, then painted over with glycerine and half dried. In order to loosen the connection of the stapes in the fenestra ovalis, a horizontal cut is made by means of a fret-saw through the pyramid, passing through the middle of the internal auditory meatus, and opening the vestibule from above without injuring the fenestra ovalis. After sufficiently widening the upper gap of the vestibule, the vestibular surface of the foot-plate of the stapes comes into view. If the upper border of the oval window be now carefully broken off with the small pointed bone forceps, and the superior edge of the foot-plate be exposed, it is easy so to loosen the lower edge of the stapes also, with the pointed preparing needle, that its connection with the fenestra ovalis is completely severed. If the ligaments of the malleus and incus be now divided, and the connection of the short process and of the manubrium with the tympanic membrane separated, the connected ossicular chain can, with careful handling, be lifted out of the tympanic cavity.

Mention might here be made of a simple manipulation by which, with a little practice, it is possible to represent the ossicular chain in its topographical relation to the superior and inner tympanic walls. For this purpose an incompletely macerated temporal bone from a new-born infant is used, in which the membrana tympani has already been destroyed by maceration, while the connection of the chain of ossicles with the tympanic walls is still a tolerably firm one. If the squamous portion of the temporal bone be first removed (see p. 24), and then the annulus tymp. carefully broken off in small pieces from above downwards, without, however, loosening the connection of the malleus within the tympanic cavity, a preparation is obtained on which, as the accompanying illustration (Fig. 75) shows, the articulated ossicular chain appears in its normal position in the cavum tymp. For the rest, as regards the study of the topographical position of the ossicles, we refer to the frontal and horizontal sections of the organ of hearing to be described in a later chapter.

Small particles of tissue adhering to the ossicles and disfiguring the preparation are, while dry, singed off by holding them close to the flame of a spirit-lamp. The separate small bones, as well as the entire chain, are fixed with cement upon a small piece of wood or glass 3 cm. high and 1½ cm. broad, which is fastened to a round stand, and protected from dust by a small glass shade (see the plate, Fig. I.,

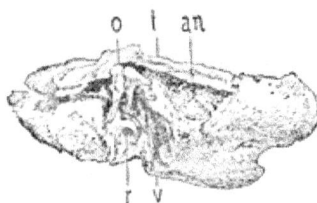

FIG. 75.—Position of the ossicles ; chain in the temporal bone of a child, after removal of the annulus tymp., and of the squamous portion.—t = tegmen tymp. ; o = ossicula audit. ; an = antrum mastoid. ; r = fenestra rotunda ; v = vagina proc. styloid.

belonging to the chapter, 'Making and preserving Normal and Pathological Preparations of the Ear for Instruction Purposes').

In pathological dissections, after opening the tympanic cavity, notice should be taken of the changes that have occurred in the ossicles. An exact examination of their exposed surfaces is only possible in those cases where the upper tympanic space is not filled with morbid products. In acute suppurative inflammations, the covering of the small bones is found loosened and reddened through considerable vascularization. In chronic adhesive processes, the mucous membrane covering, especially on the malleo-incudal articulation, appears tendinous, gray, dull, and thickened. Less frequently, there occur fine calcareous incrustations or dense calcareous deposits, appearing as though dropped upon the malleo-incudal joint. In chronic suppurative inflammations of the middle ear, the examination is far more difficult, partly on account of the swelling and proliferation of the mucous membrane covering the ossicula, partly in consequence

7

of the proliferation of the mucous membrane lining of the upper tympanic space, whereby the small bones are on all sides so enveloped by the proliferated mucous membrane that, in order to lay them bare, they require to be dissected out of the mass which surrounds them, with probe and pincette. Often enough the ossicula remain intact in the proliferated mucous membrane surrounding them; frequently, however, the head of the malleus and body of the incus are found carious, or necrosed in the midst of the proliferated mucous membrane; less frequently are the capitulum, and the crura of the stapes, destroyed. In most cases the long process of the incus perishes through fusion, and in extensive defects of the tympanic membrane this may be seen by inspection of the tympanum from the external meatus, when, on the inner tympanic wall, the capitulum of the stapes is found disconnected from the long process of the incus. The examination of the stapes in the niche of the fenestra ovalis has been already described (p. 91). The accurate investigation of the relative position of the ossicles to the newly formed connective tissue, wholly or partially filling up the tympanic space, is only possible on a series of horizontal sections of the entire decalcified organ, further details upon which will be found in the Histological part.

8. PREPARATION OF THE INTRA-TYMPANIC MUSCLES.

For the preparation of the intra-tympanic muscles, fresh objects, as well as those that have been preserved in spirits of wine, may be used. If it be intended to represent the anatomical connection of the musc. tensor tymp. with the musc. tensor veli palat., it is advisable to lay the fresh organ for eight to fourteen days in a solution of chromic acid (1 : 2000), which is to be changed several times; by the action of this the muscular fibres are sharply differentiated from the surrounding tissue.

To lay bare the musc. tensor tymp., the roof of the tympanic cavity is removed (see p. 64), and the malleo-incudal articulation and the tendon of the tensor tympani are exposed. Then an elastic probe is introduced into the osseous portion of the Eustachian tube below the tensor tendon, and with a chisel, applied obliquely, the upper wall of the canalis pro tens. tymp., from the rostrum cochleare to the place of union of the cartilaginous with the osseous Eustachian tube, is, in the direction of the latter, removed in layers. The anterior extremity of the muscle, surrounded by a fibrous sheath, arises, partly from the great wing of the sphenoid, partly from the uppermost portion of the cartilaginous Eustachian tube. Sometimes the anterior extremity of the muscle is connected by tendinous bands of fibres, or by actual muscular bundles, with the musc. tensor veli palat. (L. Mayer,

Rüdinger). In order to show this connection, the part of the great
wing of the sphenoid, roofing over the cartilaginous Eustachian tube
and adjoining the body of the sphenoid, is, together with the superior
portion of the proc. pterygoid., taken away with the chisel, and the
work completed with the scalpel. The spindle-shaped belly of the

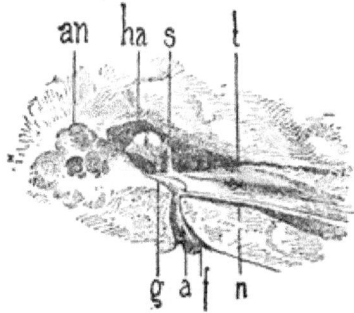

FIG. 76.—View of the tympanic cavity after removal of the tegmen tymp. (Right ear.)—
ha = malleo-incudal articulation ; t = musc. tens. tymp. ; s = tendon of the musc. tens.
tymp. passing across the tympanum ; f = nerv. facialis ; g = genu nervi facialis ; n =
nerv. petros. superf. major ; a = nerv. acusticus ; an = antrum mast. After a preparation
in my collection.

muscle receives also some muscular bundles, which spring, according
to Helmholtz, from the superior wall of the osseous canal. With a
little care the muscle, which exceeds 2 cm. in length, may be successfully raised out of the canal with a small knife rounded at the point ;
and after detaching its tendon at the rostrum cochleare, it may be
isolated and represented in connection with the malleus.

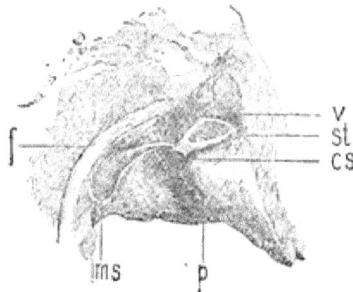

FIG. 77.—Posterior portion of the inner tympanic wall (Right ear). Enlarged to double its
size.—st = stapes ; cs = capitulum stapedis ; ms = musc. stapedius in the cavitas
stapedii, with its tendon inserted at the capitulum ; p = promontory ; f = nervus facialis ;
v = vestibule laid open. After a preparation in my collection.

The musc. stapedius is most easily represented after separation of
the pyramid from the pars tymp. This small, pear-shaped muscle
(Fig. 77, st) is enclosed by the already mentioned eminentia pyramidalis, projecting at the inner part of the posterior wall of the
tympanic cavity. At the point of the emin. pyramidalis, directed

7—2

forwards and upwards, its tendon passes to the capitulum of the stapes. The length of this muscle varies between 6 and 7 mm.

The musc. stapedius is exposed in the simplest way by carefully removing the outer osseous lamella of the eminentia pyramidalis, the position of which is given in the prolongation backwards and downwards of the tendon of the stapedius. For this purpose, either a narrow, sharp hand-chisel or a broad graver (Fig. 12, a) is made use of, with which, by moderate pressure, the osseous lamella covering the muscle can be broken off. The exposed muscle is to be lifted out of its bony groove by means of a narrow, small knife rounded at the point, and may be preserved, in connection with the stapes, in spirits of wine.

In order to make the intra-tympanic muscles easily distinguishable on preparations which have been preserved in spirits of wine, and which are intended for demonstration purposes, they are brushed over with a solution of carmine, in doing which, care is to be taken that the colouring fluid does not spread to the surrounding tissues.

In pathological dissections, one should never fail to examine the intra-tympanic muscles. With the musc. tensor tymp. this is possible under all circumstances; with the musc. stapedius only when either the dissection permits of separation of the membrana tympani from the pars petrosa, or when, while leaving these two in connection, the tympanic cavity is opened from behind after the method suggested by Lucae (p. 70). In the latter case the facial nerve, lying behind and laterally from the muscle, has to be removed in order to expose the belly of the muscle. In non-purulent middle ear processes, pathological changes in the muscles can only rarely, and then only in advanced atrophy, be demonstrated macroscopically, for which reason small portions of the belly of the muscle should always be taken from the preparation for microscopic examination. More frequently one finds in purulent middle ear processes on the intra-tympanic muscles, appearances of hyperæmia, discoloration, and, in carious processes, erosion of the osseous sheaths of the muscles, and of the muscles themselves, as also destruction of their tendons. Where, in pathological dissections, the entire organ is to be subjected to microscopic examination, and where, at the same time, the changes in the intra-tympanic muscles are to be studied, in making the cut, regard should always be had to the direction of the course of the muscles.

9. PREPARATION OF THE CARTILAGINO-MEMBRANOUS EUSTACHIAN TUBE AND ITS MUSCLES.

The opening of the osseous Eustachian tube from the side of the tympanic cavity, after removal of the musc. tensor tymp., has been previously described (see p. 64).   The preparation of the cartilagino-membranous Eustachian tube requires an accurate knowledge of its situation on the lower surface of the basis cranii.   The cartilaginous structure of the tube, from its firm fibro-cartilaginous union with the osseous tube to the ostium pharyng. tubæ, lies in a groove known as the sulcus tubæ Eustachii on the lower surface of the base of the skull. This groove, lying between the pyramid of the petrous bone and the great wing of the sphenoid, passes forwards as a furrow on to the medial lamella of the proc. pterygoid., to terminate at its posterior edge, somewhat above the middle of the process.   The cartilaginous portion of the Eustachian tube is fixed in this groove in such a way that the roof and the medial plate in the upper portion of the tube are connected partly with the groove, partly with the fibrous tissue which closes the fissura spheno-petrosa.   At this spot the cartilage of the tube is firmly fixed, and almost immovable.   Lower down, however, in the vicinity of the proc. pterygoid., the medial cartilaginous plate, projecting coulisse-like at the lateral pharyngeal wall, appears to the extent of 1-1½ cm. easily movable.   The anterior edge of this projecting part of the medial plate forms, with the thick fold of mucous membrane (plica salpingo-pharyngea) passing downwards from it, the posterior prominent portion of the tube.   The reverse is the case, as regards mobility, with the lateral plate of the cartilaginous tube, known as the cartilage hook.   While the major part of its posterior portion shows greater mobility than the medial plate, the anterior inferior extremity is fixed, by dense connective tissue, with the outer wall of the sulcus tubæ, and is only slightly movable.

It will be best to enter upon the representation of the anatomical relations of the Eustachian tube by commencing with the preparation of its muscles.   In order to lay bare the levator palati mollis (Fig. 78, 11'), the mucous membrane is laid hold of with a pincette in a thin layer ½ cm. below the orifice of the Eustachian tube (os) in a skull which has been divided into halves in a sagittal direction, then raised and cut away with curved scissors.   When the muscle comes into view through the incision in the mucous membrane, the direction of the course in which the muscular bundles run appears so sharply marked that, guided by them, it is possible to expose the entire muscle, from its origin at the lower surface of the point of the pyramid and the contiguous part of the tubal cartilage (l), up to its radiation in the velum palati (l').

If now, from the opening made below the ostium tymp., the mucous membrane of the velum palati be dissected away in the direction of the fibres of the belly of the muscle, it is possible to accurately trace the radiation of the muscular bundles on the posterior surface of the velum (l'). The greater number of them lie in the middle, between the upper edge of the velum and the tip of the uvula. When the lower portion of the muscle has been exposed, it is possible, after splitting the mucous membrane below the median cartilaginous plate, to isolate also the upper portion of the levator by loosening, with the narrow handle of a scalpel, the slight connection of the muscle with the membranous floor of the Eustachian tube up to its origin at the base of the skull.

By the levator palat. mollis, in consequence of the shortening and

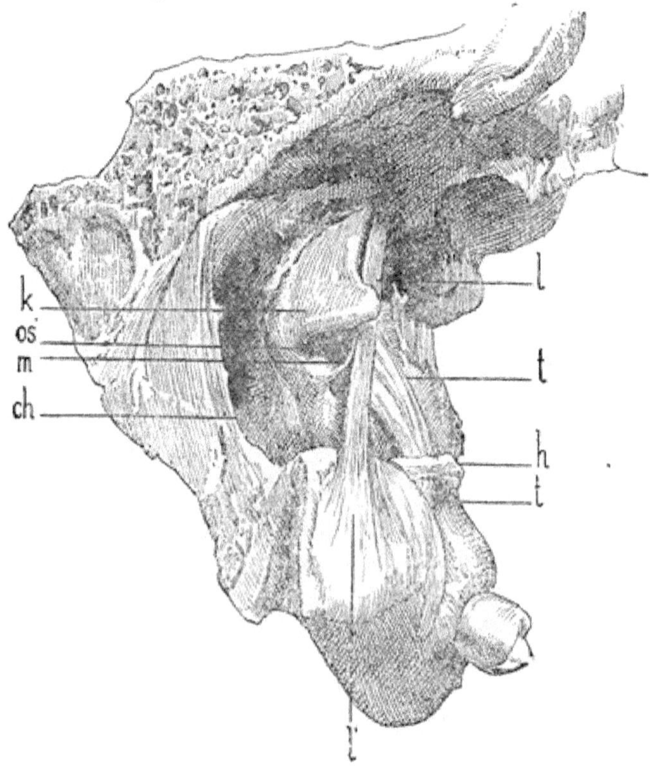

Fig. 78.—Eustachian tube with its muscles; natural size. (Right side.)—k = cartilaginous plate of the Eustachian tube; m = pars membranacea of the Eustachian tube; os = ostium pharyng. tubæ; ch = choana; l = musc. levator palati mollis; l' = radiation of the levator in the velum palati; h = hamulus pterygoideus; tt = musc. tensor palati mollis, winding round the hamulus. After a preparation in my collection.

swelling of its belly, not only is the floor of the Eustachian tube raised up, but the median cartilaginous plate is also pressed considerably inwards, whereby the ost. pharyng. tubæ is certainly diminished in its high diameter; but the resistance in the tubal canal itself is lessened by the widening of the slit of the tube. It is beyond all doubt that the action of the tensor veli palat., as a dilator of the canal of the tube, is materially assisted by the co-operation of the levator veli palati.

Somewhat more difficult is the preparation of the musc. tensor veli palat. (dilatator tubæ, abductor tubæ, Von Tröltsch). This muscle arises partly from the inferior surface of the wing of the sphenoid medially from the foramen ovale et spinosum, along a line extending forwards as far as the proc. pterygoid., partly, and with the great mass of its bundles, from the whole length of the lower margin of the lateral cartilaginous hook, and from the lateral membranous part of the cartilaginous section of the tube.

The preparation and laying bare of this muscle is effected in part from the naso-pharynx, and in part from the lateral side of the tube. The work is materially facilitated if the divided skull be previously so reduced in size that the muscle can be easily reached from its lateral side. For this purpose the superior maxilla is sawn away 1-2 cm. in front of the choanæ in a frontal direction, and in addition to this the preparation is reduced at its posterior and outer portion, by removing the os occipitale and the squamous portion of the temporal bone. At the same time the bones and soft parts adjoining the lateral side of the Eustachian tube are removed by cutting out a wedge-shaped section on the external surface of the skull. The posterior cut of this wedge passes 1 cm. in front of the anterior wall of the meatus, and parallel to it, through the glenoid cavity as far as the inner part of the fissura orbitalis superior; while the anterior one separates the still remaining portions of the zygomatic process and of the superior maxilla, and meets the former cut at the fissura orbitalis superior.

On the preparation, thus reduced in size, the levator veli palat. (Fig. 78, l l'), already prepared, is next lifted up from the side of the naso-pharynx, and the lower portion of the musc. tens. veli palat. (Fig. 78, t t), now visible, is so far exposed on that side as the Eustachian tube, covering it in part, and the levator veli palat. will allow.

To expose the muscle from its lateral side also, the remainder of the articular capsule of the inferior maxilla and of the musc. pterygoid. extern. in the space left on the outer surface of the preparation by the removal of the wedge, is cut away up to a point where the external lamella of the proc. pterygoid. comes into view. If we now break off the latter with the pointed bone forceps from its lower edge up to the foramen ovale, and afterwards remove the musc. pterygoid. intern. with pincette and scalpel, we obtain a complete side-view of the musc. tensor veli palat., with its origin at the sphenoid bone and at the cartilaginous tube, as also the transition of the muscle (Fig. 78, t t) into its tendon, winding round the hamulus pterygoid. (h), and radiating into the soft palate.

The tensor veli, the tendon of which radiates in the fibrous prolongation of the hard palate, is the true dilator of the Eustachian tube, as by its contractions the hook of the tubal cartilage is rolled up, and the membranous portion of the tube is drawn away from the cartilaginous (Von Tröltsch).

A general view of the muscular apparatus of the tube on both sides, and its relation to the soft palate, is obtained by dissection from behind. We proceed in the following manner: By means of a frontal cut with the saw through the skull, the occiput is removed as far as the tuberculum pharyng., and the posterior membranous pharyngeal wall cautiously exposed. On the intact pharynx it is now very easy to prepare the superior portion of the constrict. pharyngis super., as well as the upper part of the levat. veli palat., and laterally from this the tensor veli palat. In order to show also the lower portions of both muscles and their radiation in the velum palati, the posterior pharyngeal wall is split in its middle by a vertical cut, and the preparation completed from the opened naso-pharynx.

The preparation of the cartilaginous Eustachian tube itself is, after what has been said, included in that of the muscles of the tube, since, in exposing the latter, the greatest part of the median and lateral wall of the tube is laid bare. The muscles and cartilagino-membranous portion of the Eustachian tube having been exposed on a preparation in the manner before described, it is necessary, in order to expose the roof of the cartilaginous and its junction with the osseous portion, to divide first, with a narrow scalpel, in the direction from the naso-pharyngeal space, the attachment of the cartilaginous roof of the tube at the base of the skull, and to remove, from this incision, the body, along with the great wing of the sphenoid bone by means of the fret-saw. If, now, any particles of the sphenoid, left projecting over the tubal cartilage, are broken off with the forceps, a complete representation of the position of the entire Eustachian tube and of its muscles is obtained.

As regards the method of opening the tubal canal, we must, in order to avoid repetition, refer to the previous description (pp. 71 and 72). The splitting of the cartilaginous roof of the tube, from the ost. pharyngeum tubæ to the isthmus, is best effected with a narrow, small knife, or with scissors, taking as a guide a metal probe pushed forward into the osseous tube, which has been previously opened with the chisel from the side of the tympanic cavity.

The representation of topographical preparations of the Eustachian tube is to be found under the heading 'Representation of Topographical Preparations of the Organ of Hearing.'

The insertion of the cartilaginous portion into the rough parts around the osseous tube extends laterally higher up towards the tympanic cavity, than it does in a median direction. The axes of the osseous and cartilagino-membranous portions do not lie exactly in the same line, but form an obtuse angle open downwards (p. 72, Fig. 62), corresponding to the isthmus tubæ. According to Henle the canal of the tube occupies a position nearly exactly diagonal between the transverse and the sagittal. The angle open in front, which the axis of the tube forms with that of the external meatus, measures, according to Schwalbe, 150°; with the horizontal, the axis of the tube forms an angle of 40° (Henle).

The more exact anatomical relations of the cartilagino-membranous Eustachian tube will be discussed in the Histological part.

In pathological dissections, where the principal object is to ascertain morbid changes of the mucous membrane of the tubal canal, the longitudinal splitting of the cartilaginous portion and the simultaneous opening of the osseous tube is absolutely necessary. The proceeding corresponds exactly with that of the opening of the canal of the tube in the normal organ.

Attention is here to be directed to the accumulation of purulent or mucous secretions in the canal of the tube, to the degree of hyperæmia and swelling of its mucous membrane, to the colour of the same, to the presence of small ulcers and cicatrices (most frequently in the neighbourhood of the ostium pharyngeum tubæ), and to granulations and bridges of connective tissue in the tubal canal. Where, during life, a stricture of the tube had been diagnosed, it is better not to split the tubal canal, but to subject the cartilagino-membranous portion, removed along with the osseous part of the tube, to histological examination, because the degree of the stricture, and the alterations in the tissue resulting therefrom, can thus be more thoroughly investigated than is possible by anatomical preparation and the splitting of the canal. In like manner, neoplasms encroaching on the canal of the tube from the neighbouring structures, such as carcinoma, sarcoma, or cyst formations in the vicinity of the canal, and diverticular formations in the Eustachian tube (Kirchner), are better examined on microscopic preparations.

10. PREPARATION OF THE MASTOID PROCESS.

To study the anatomical relations of the mastoid process, which in the adult, as regards size, form, and internal structure, presents many varieties, a considerable number of macerated temporal bones will be required.   With the preparation of sections on dry temporal bones the comparative examination on fresh specimens must always go hand-in-hand.

First, to become acquainted with the internal structure of the mastoid process, a cut nearly parallel to the external surface is made with the fret-saw, from the lower point of the mastoid process to the level of the superior wall of the osseous meatus, by which about a third of the process is removed (Figs. 79 and 80).

Its internal structure varies so much that even in one and the same individual the interior of the one side hardly ever resembles that of the other.

Although some of the older authors had noticed that the mastoid process does not always consist entirely of pneumatic cell spaces, but frequently wholly or in part of adipose, diploëtic or sclerotic

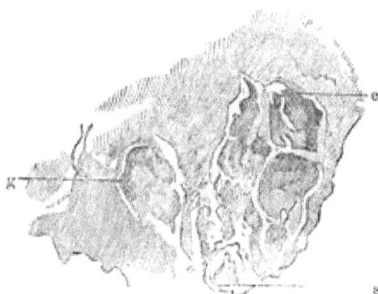

FIG. 79.—Pneumatic mastoid process.   FIG. 80.—Diploëtic mastoid process.

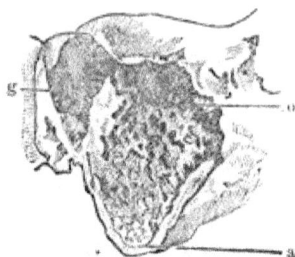

osseous substance, it was Zuckerkandl who, after numerous investigations, first pointed out that only in 36.8% of the cases the mastoid (Fig. 79) consisted, from its lowest point to its upper boundary (g), of larger or smaller pneumatic spaces, but that in 43.2%, it was formed, chiefly in its lower (Fig. 82) and posterior (Fig. 81, d' d) portion, of diploëtic osseous substance, while the anterior superior part (Fig. 81, d' p) contained pneumatic cell spaces; and, finally, in 20% the mastoid process was wholly made up of diploëtic (Fig. 80), less frequently of sclerotic tissue.

For the study of the anatomical relations which, in operations on the mastoid process, are chiefly to be considered, it is necessary to make a considerable number of sections of the temporal bone, on which the position of the antrum mastoid. to the sinus transversus and to the external surface of the mastoid process is brought under observation.

As to the extent of the antrum, a general view may be obtained if, after the method previously alluded to (p. 64), the tegmen antr.

mastoid. be removed along with the tegmen tymp. with chisel or forceps. A glance from above enables us to see the anterior boundary of the antrum, its extent backwards, its depth, as well as the size and number of the openings by which it communicates with the mastoid cells.

A correct impression of the greatly variable relative distances of the antrum to the external meatus, to the sinus transversus and to the cortical substance of the mastoid process, is, however, only obtained on sections which are made through the process in a horizontal and in a vertical direction.

The horizontal cut through the temporal bone, which is to expose the antrum mastoid. in its greatest transverse diameter, commences on the posterior surface of the petrous bone at the porus acustic.

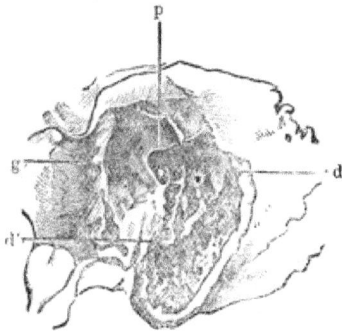

FIG. 81.—Diploëtic mastoid process with few air-containing cell spaces in its anterior upper portion.

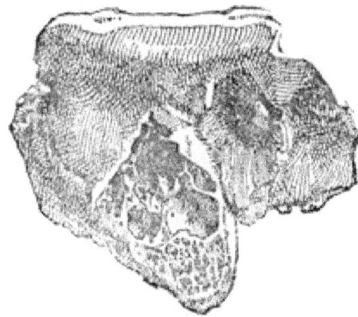

FIG. 82.—Mastoid process, with a pneumatic structure in its upper, and a diploëtic structure in its lower portion.

intern., passes in a horizontal direction through the upper tympanic space, and terminates at the external surface of the temporal bone, immediately above the upper periphery of the external opening of the osseous meatus.

Such sections are suitable for exact measurement of the distances: from the inner boundary of the antrum to the sinus transversus; from the external wall of the antrum to the external surface of the mastoid process; and from the sinus transversus to the posterior wall of the meatus.

As regards the relative position of the antrum to the sinus transversus, these are frequently enough found separated only by the thin, but compact, rarely dehiscent lamella of the osseous wall of the sinus; oftener, however, antrum and sinus are separated by a broad layer of pneumatic (Fig. 47, p. 45) or diploëtic cellular spaces, which latter afford a certain protection against the spreading of a purulent inflammation from the mastoid process to the sinus. In the new-born

infant there lies constantly, between the antrum and the sin.
transvers., a broad layer of diploë (Zuckerkandl), which, as the
accompanying illustration (Fig. 83) shows, is generally to be met
with in the first few years of life.

Measurements on horizontal sections of the temporal bone are of
importance for the operation of opening the mastoid process, in order
to determine the distance between the external wall of the antrum
and the field of operation, situated in the anterior superior part of
the outer surface of the mastoid process. Its upper boundary lies
almost on a level with the superior osseous wall of the meatus; the
anterior boundary at the point of transition of the planum mastoid.
into the posterior wall of the osseous meatus. Numerous measurements on horizontal sections made by me showed that the distance
from this point to the antrum varies between 6 and 15 mm.

Still more important is the study of the individually greatly

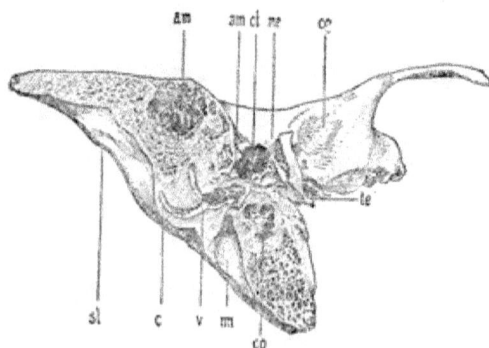

Fig. 83.—Horizontal section through the temporal bone of a child one year old.   Upper half
of the section.—me = meatus audit. extern. ; a = anterior wall of the meatus ; ct =
cavum tympani ; am = antrum mastoid. ; cg = cavitas glenoidalis ; mi = meat. audit.
intern. ; co = cochlea ; v = vestibule ; c = canalis semicircul. horizontal. ; st = sinus transversus.

varying relative positions of the sinus transversus to the posterior
wall of the meatus and to the external surface of the mastoid process.
As a rule, especially with chiefly pneumatic processes, the sinus, the
posterior wall of the meatus, and the planum mastoid., are separated
by a broad interspace formed of cellular spaces (Fig. 84), through
which, in opening the mastoid process, it is easy to penetrate as far as
the antrum mastoid. without incurring the risk of opening the sinus.
The osseous groove of the latter appears in such preparations flat,
very little curved, and only to a slight degree excavated at its
superior part, which is bridged over by the posterior continuation of
the upper edge of the pyramid. In other, but certainly much less
frequent cases, the groove of the sinus is found considerably

deepened, strongly curved, and in its middle and upper portions arched so far outwards and forwards that between it, on the one side, and the posterior wall of the meatus, as well as the external surface

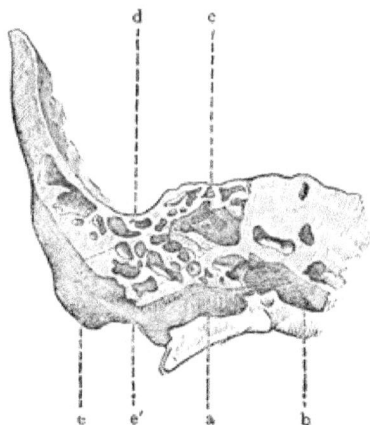

FIG. 84.—Horizontal section through a pneumatic mastoid process.—a = posterior wall of the auditory meatus ; b = tympanic cavity ; c = antrum mastoid. ; d = sin. transv. ; e e′ = place for operation on the outer plate of the mastoid process.

of the mastoid process, on the other side, merely a narrow bridge of bone is present (Figs. 85 and 86). It is evident that in such cases the opening of the mastoid process is attended with great danger, since, while advancing towards the antrum, one may expose and

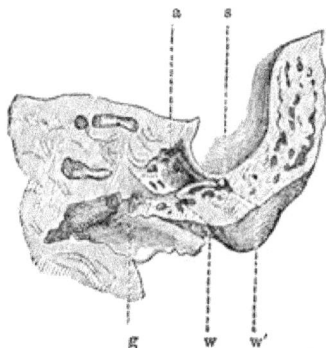

FIG. 85.—Horizontal section through a partly diploetic, partly pneumatic mastoid process.—g = posterior wall of the auditory meatus ; a = antr. mast. ; s = sin. trans. ; w w′ = seat of operation.

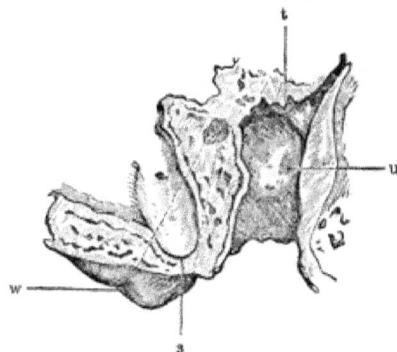

FIG. 86.—Horizontal section through a compact mastoid process with but few diploetic spaces. —t = tympanic cavity ; u = lower wall of the meatus ; s = sin. transv. ; w = posterior boundary of the seat of operation.

injure the sinus. Having made numerous sections of temporal bones, I was the first to point out that this relation of the sinus, so unfavourable for operating, is chiefly met with in temporal bones with a diploëtic or sclerotic structure.

For the purpose of representing in one and the same temporal bone the position of the antrum mastoid. and the structure of the mastoid process, after previously exposing the mastoid cells by means of the vertical cut suggested on p. 106 (Fig. 87, ma), we carry parallel to the upper surface of the pyramid a cut with the saw, describing a curve with its concavity downwards (Fig. 87 s s′ s″), which commences in front of the pars squamosa (s), and runs towards the posterior edge of the temporal bone in such a way that it meets the root of the zygomatic process, the upper wall of the osseous meatus, and the uppermost part of the antrum mastoid. The separated

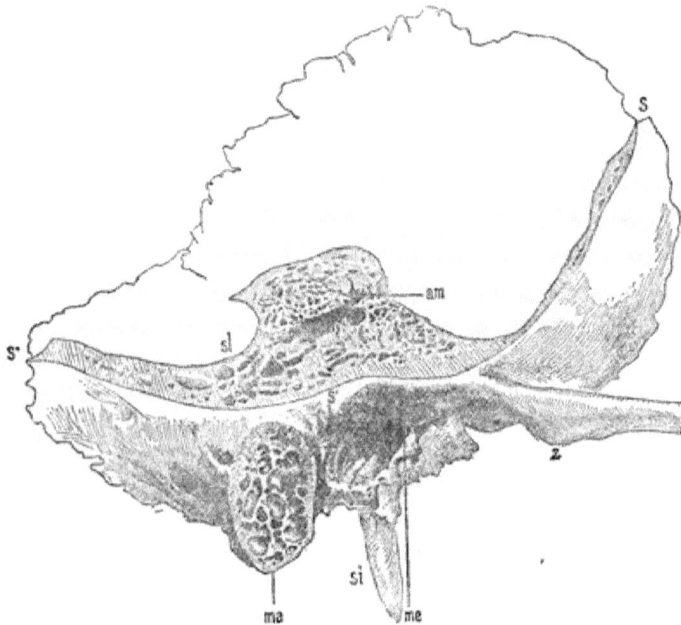

Fig. 87.—Curved section through the temporal bone.—me = meatus auditor. extern. ; ma = section of a pneumatic mastoid process ; s s′ s″ = curved section through the temporal bone ; am = antrum mastoid. ; sl = sinus transversus ; z = processus zygomaticus ; st = processus styloideus.

parts of the temporal bone should be so connected with brass wire and small brass hooks, that the posterior part, situated above the pars mastoid. (s″), may be moved horizontally towards the lower part of the preparation, by which means a general view of the position of the antrum and its relations to the neighbouring parts is obtained.

In addition to the sections of the temporal bone so far mentioned, it is necessary, in order to become familiar with the anatomical relations of the mastoid process, to prepare a number of vertical sections, both in a frontal and a sagittal direction.   Successful sagittal sections (Fig. 88), which pass through the antrum and the tip of the mastoid process, give a good general idea of the length and depth of the antrum, and of the arrangement of the mastoid

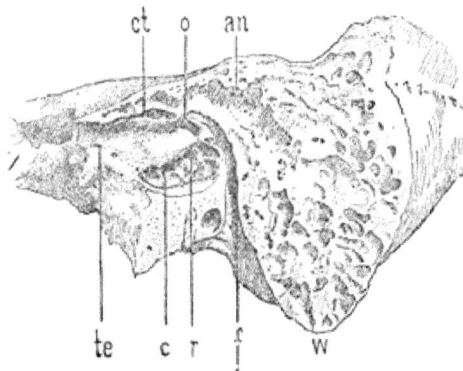

Fig. 88.—Sagittal section through the middle of the tegm. tymp. and lower tympanic wall : lateral view of the inner half. (Left ear.)—te = tub. Eust. oss. ; ct = canalis pro tens. tymp. ; c = lower ridgy wall of the tympanic cavity ; o = fenestra ovalis ; r = fenestra rotunda ; an = antrum mastoid. ; f = canalis Fallopii, communicating towards the front with the cavit. stapedii ; w = mastoid process.

cells opening into it. On frontal sections, moreover, particularly with pneumatic mastoid processes, the relation of the mastoid cells to the

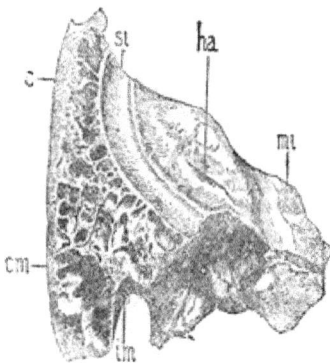

Fig. 89.—Frontal section through a pneumatic mastoid process and the incisura mastoidea.—c cm = corticalis of the pneumatic mastoid process ; im = incisura mastoidea ; si = sinus transversus ; ha = hiatus aquæd. vestibul. ; mi = porus acust. internus.

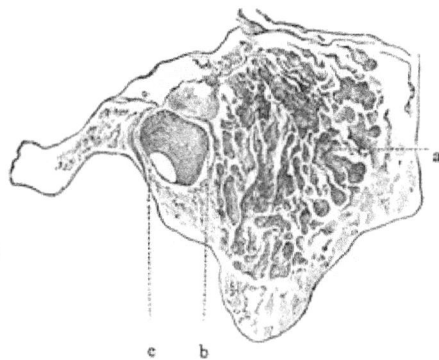

Fig. 90.—Vertical (sagittal) section through the mastoid process and osseous meatus.—a = mastoid cells ; b = posterior wall of the osseous meatus ; c = anterior wall of the osseous meatus.

incisura mastoidea is to be seen (Fig. 89). Here are not infrequently found on the section one or several cyst-like air-spaces, the wall of which, thin as paper, borders immediately on the incisura mastoidea.

Accumulation of pus in these cavities may, in some cases, lead to rupture of the incisura and to sub-muscular and sub-fascial burrowing of pus in the neck (Bezold).

Other sagittal and horizontal sections made through the mastoid process and the osseous meatus give an instructive representation of the position of the mastoid cells lying laterally from the antrum mastoid., and of their relation to the posterior superior wall of the osseous meatus. Sagittal sections (Fig. 90) which are to demonstrate the relation of the mastoid cells to the osseous meatus must be made parallel to the longitudinal axis of the pyramid, in a plane which passes through the tip of the mastoid process, through the external portion of the osseous meatus, and through the inner part of the glenoid cavity. Horizontal cuts (Fig. 91), carried through the axis of the osseous meatus, best show the extent of the mastoid cells behind the meatus in a lateral direction. That the number and

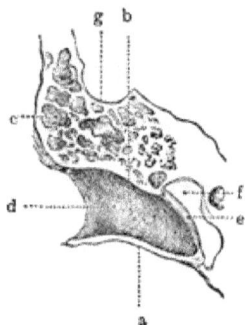

FIG. 91.—Horizontal section through the external meatus and mastoid cells. (Right ear.) —a = anterior wall of the meatus ; b = its posterior wall ; c = mastoid cells ; d = auditory canal ; e = tympanic membrane ; f = tympanic cavity ; g = sinus transversus.

arrangement of these cellular spaces stand in a certain relation to the pneumatic or diploëtic condition of the mastoid process has already been stated (pp. 33-36). (Compare also the chapter 'Sections of the Temporal Bone of the Adult,' pp. 33-47.)

The anatomical relations of the antr. mastoid. in the new-born infant were discussed in the dissection of the temporal bone of the new-born infant (pp. 27-29).

In the anatomical dissection of the processus mastoid. on fresh unmacerated preparations, the proceeding, on the whole, is the same as that with macerated temporal bones. That in the normal organ the antrum, opened from above, frequently shows fine, knotty filaments, with those pedunculated, stratified formations first described by me, as well as serrated membranes, which are inserted, with delicate off-shoots, on the walls of the antrum, has already been mentioned. On longitudinal sections of the processus mastoid., the morphological differences between pneumatic and diploëtic structure are even more marked than on the macerated temporal bone. The pneumatic cellular spaces are large, pale, and filled with air, while the diploëtic are small, and filled with a reddish or yellowish-red substance, and, on the surface of the cut, present the appearance of a section of a cylindrical bone.

The preparation of the mastoid process in pathological cases depends upon the seat and extent of its morbid condition. A thorough examination of the proc. mastoid. is particularly required in purulent middle ear inflammations and their consequences, and even if, during life, no symptoms of any affection of the mastoid process had existed. In acute, purulent, middle ear inflammations, the antrum and the pneumatic cells are, without exception, filled with pus or muco-purulent secretion; the lining membrane of the cellular spaces is reddened and tumefied. In chronic middle ear suppurations, the cellular spaces are also frequently filled with pus. The mucous membrane lining of the mastoid cells appears dark red, tumefied, pro-liferating, or degenerated by polypi (Trautmann). Accordingly, while the suppuration lasts, but mostly after it has ceased (much less fre-quently in non-purulent catarrhs), the antrum and mastoid cells are found to be completely filled with newly-formed, gelatinous or fibrous connective tissue. Should such proliferations of connective tissue, after having existed a considerable time, become ossified, the whole of the mastoid process is converted into a uniform mass of bone, in which, when cut with the saw into sections, the cellular spaces are either entirely wanting, or present only in small numbers. (Eburna-tion of the mastoid process.)

In caries and necrosis of the mastoid process, after previously opening the antrum from above, the cortical portion of the external surface of the process is opened with chisel and forceps, in order to ascertain the extent of the disease in the bone. If the outer bony shell be perforated by a fistulous opening, the interior of the process. mastoid. should be laid bare by widening the already existing aperture in the bone by means of the forceps, and by careful sounding it should be determined whether a movable or firmly fixed sequestrum of the bone is present, and whether, after its removal, the osseous walls of the cavity feel rough, or are already covered over with smooth granulation tissue. At the same time, by the use of the probe, the direction and length of any fistulous canals between the mastoid process and the external meatus or the sinus transversus are determined, and in multiple fistulous formations in the neigh-bourhood of the ear, their connection with the mastoid process,

8

or with other parts of the temporal bone, is made clear by anatomical preparation and exposure of the fistulous passages.

As in carious-necrotic processes, we proceed also with the so-called cholesteatomatous formations in the mastoid process which, as white, or whitish yellow tumours, iridescent on the surface, may sometimes be taken out of the smooth tendinous, gray cicatricial tissue-lining of the cavity of the process. More frequently there are found in the latter shapeless caseous masses permeated by the detritus of sequestra, simultaneously with retention products in the tympanum. After removal of these masses, the temporal bone very often presents a system of cavities and sinuses, which extends beyond the boundaries of the mastoid process, and is caused by partial fusion of the cavities of the process, of the tympanum and of the external meatus.

11. PREPARATION OF THE LABYRINTH.

(a) *Preparation of the Osseous Labyrinth.*

The preparation of the osseous labyrinth is much more easily carried out on the petrous bone of the new-born infant than on that of the adult, because the superjacent bone surrounding the osseous labyrinth in the one case is still porous, soft, and only loosely connected with the labyrinthine capsule, while in the other the connection of the labyrinth with the solid mass of bone surrounding it is so complete that the boundaries between the capsule and the overlying bone entirely disappear. On the petrous bone of the new-born infant, therefore, it is possible, without special previous practice, by careful scraping of the porous, overlying bone with a short, sharp knife, to prepare out the osseous labyrinth in its outlines.*

In the petrous bone of the infant it is best to begin by laying bare the semicircular canals, first of all the superior one. Its position is marked by the eminentia arcuata, and below this by a depression ending in a cul-de-sac, and called fossa subarcuata (Fig. 21, p. 24). First, the spongy and soft osseous mass lying behind this canal is removed in layers with a knife, on the one side up to the point where the superior and posterior canals meet, and on the other side up to the compact wall of the horizontal canal. After this, the exposure of the posterior sagittal canal is proceeded with. Its upper crus projects on the posterior surface of the pyramid, and can therefore easily be laid bare. The curve and the inferior crus of this canal are, however, surrounded by a thick layer of the overlying bone, which must be carefully removed to make the crus free. The preparation of the horizontal canal, the

---

* According to Ilg and Hyrtl, the proceeding is considerably facilitated by first boiling the petrous bone for an hour in a solution of caustic potash ; this treatment, however, gives the preparations a yellow colour.

outer crus of which projects on the inner wall of the tympanic
cavity behind the fenestra ovalis, is effected by removing in layers the
osseous mass still remaining behind and below, the difference between
the porous structure of the bony covering, and the compactness of the
walls of the canals, serving as a valuable guide while proceeding with
the preparation.

When the semicircular canals have been exposed in their rough
outlines, the finer work of preparing commences. In the first place,
the space below the superior canal is made accessible by piercing,
with a straight, sharp, awl-like instrument or drill, the remaining mass of bone. Then a thin, round file is introduced into the bore-
hole, and the remainder of the overlying bone is filed away up to the
concavity of the canal.

The isolation of the posterior semicircular canal is more difficult,
as the inner crus of the horizontal crosses the posterior canal and partly
projects into the space within the concavity of the latter. In order,

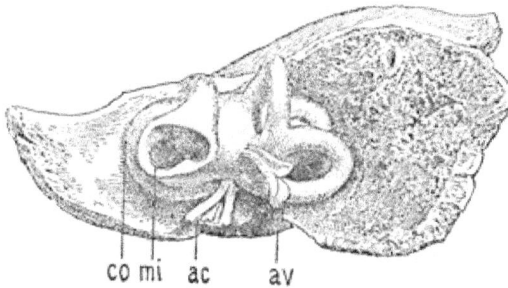

FIG. 92.—Posterior view of the osseous labyrinth, with the three semicircular canals, the
cochlea and the aquæducts of a new-born infant. (Enlarged to double its size.)—co =
cochlea ; mi = meat. audit. intern. ; ac = aquæductus cochleæ ; av = aquæductus vestibuli.

therefore, to free the posterior canal, the osseous substance between
the inner crus of the horizontal, and both crura of the posterior canal,
is so far cut out with the point of a penknife as to obtain two slit-like
openings, which, being widened by a narrow, flat file, isolate the two
canals one from the other. The particles of bone remaining between
them are removed partly with the sharp point of the knife, partly
with the file. After this, the soft osseous substance lying within the
horizontal canal is pierced vertically with a drill or pointed, awl-like
instrument, and this canal likewise cleared by means of a round file.
Finally, by careful filing and scraping, with suitable instruments (Fig.
13), of any remaining roughness on the surface of the canals, the preparation of the posterior portion of the osseous labyrinth is completed.

In order to represent the capsule of the cochlea in the new-born
infant, the soft, superjacent bone is removed with a short sharp knife

8—2

either from the side of the labyrinth or from the point of the pyramid, until the inferior cochlear convolution is reached. In this manipulation it is all the more easy to recognise the cochlear capsule, as the space intervening between it and the superjacent bone is large-celled and porous, and the gray capsule glimmers through as a uniform and more compact structure than the spongy bone over it. Yet the preparation of the capsule of the cochlea requires considerable care, because its compact osseous substance is thin as paper and very brittle, and breaks on the slightest pressure. Inequalities on the exposed semicircular canals and the cochlea are removed by careful rubbing with fine emery-paper.

On the pars petrosa of the embryo (6-8 months), in which the labyrinthine capsule is surrounded merely by a thin layer of bone, the osseous labyrinth can be much more easily dissected out than in the temporal bone of the new-born infant.

Much greater are the difficulties met with in the preparation

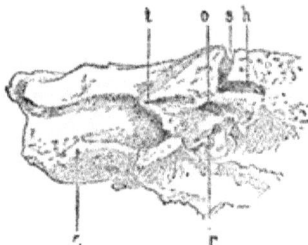

Fig. 93.—Lateral wall of the pyramid.—o = fenestra ovalis ; r = fenestra rotunda; t = semicanalis pro tens. tymp. ; c = canalis caroticus ; s = canalis semicirc. superior ; h = canalis semicirc. horizontalis.

of the osseous labyrinth in the adult, the capsule and the overlying bone being here fused into an exceedingly dense mass, which is no longer anatomically separable. Even if we succeed in preparing out the form of the osseous labyrinth from this extremely hard mass of bone, we must not by any means imagine that we have the actual capsule. Although this preparation, in anatomical respects, possesses much less value than the morphological representation of the labyrinth by corrosion, yet, for the sake of completeness, we did not feel justified in entirely passing over this method of preparing the labyrinth in the adult.

The boundary of the capsule not being traceable in the mass of the petrous bone of the adult, the semicircular canals must be opened in order to obtain a guide for preparing it out. For this purpose, the osseous mass bordering posteriorly on the canals is first sawn away at a distance of 1 cm. behind the elevation of the superior canal. Then, by filing off the eminentia arcuata, the canal is opened at its convex curvature. The slit now appearing to view stands vertical to the axis of the pyramid, and gives us the direction of the superior canal, which from this point can be easily opened in its entire length, by filing off, in the direction of the slit, the osseous mass in an outward direction up to the ampullary dilatation on the one side (Fig. 93), and inwards as far as the union with the posterior canal on the other side (Fig. 94).

The preparation is now so fixed in the vice, that the posterior surface of the pyramid faces upwards. If then the surface of bone lying behind the p or. acust. int. be filed away, first the superior crus, next the arch, and lastly the inferior crus of the opened posterior canal (Fig. 94, h), come into view.

Next follows the opening of the horizontal semicircular canal,

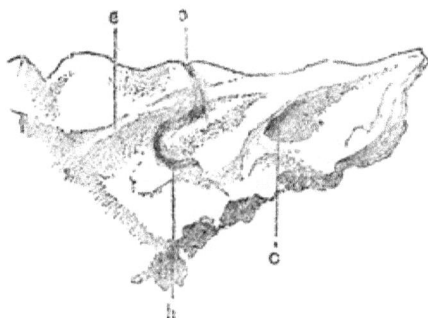

Fig. 94.—Posterior wall of the pyramid.—c = meatus audit. int. ; o = canalis semicirc. superior ; h = canalis semicirc. posterior ; e = sinus petrosus superior.

the preparation of which offers the greatest difficulties. If the object be fixed in the vice, with the posterior extremity of the pyramid directed upwards, the canalis horizontalis is seen after filing off the osseous mass lying laterally from the curve of the posterior canal. By continuing the filing in the direction towards the inner wall of the tympanic cavity, the external crus (Fig. 93, h) of this canal is also opened as far as its embouchure into the vestibule. The laying bare of the inner crus is not effected until later.

When the semicircular canals have been opened in the manner described, the osseous mass between them must be removed for the purpose of clearing them further. In order to take away the greater portion of the bone between the superior and horizontal canals, the preparation is again so fixed in the vice that the upper surface of the pyramid is directed upwards. A fine fret-saw is now applied perpendicularly, close behind the superior canal, which has already been opened, and the compact bone is sawn through, but only as far as the union of the superior and posterior canals. The saw is next applied horizontally immediately above the horizontal canal, and the cut carried up to the former vertical one, when the piece of bone between the superior and horizontal canals will fall out.

By a second cut of the saw made perpendicularly downwards in front of the superior canal, and extending about 2 mm. deep to the vicinity of the vestibule, and below gradually assuming a horizontal direction forward, the anterior side of the superior canal is likewise freed from the adjoining osseous mass. The bone, still remaining below the horizontal canal, is taken away with files and graver.

The removal of the osseous substance within the concavity of the canals is effected by means of a drill, piercing first the surface of the bone below the superior, and then that within the horizontal canal. The bore-hole is now widened by using larger round files in succession, until so much of the bone is filed away that only a thin lamella remains below the opened canals.

Most difficult is the clearing of the concave side of the posterior and of the inner crus of the horizontal canal. This is most quickly accomplished if the osseous mass, lying within the posterior canal, be pierced in two places, close below the superior, and above the lower crus, with a very fine drill, in the direction towards the inner wall of the tympanic cavity. Between the two bore-holes there still remains in the osseous mass the inner crus of the horizontal canal.

With thin, round files, and afterwards with a narrow, fine, flat file, these two openings are gradually lengthened, and, making special use of the latter instrument, the bone lying between the arch of the posterior and the inner crus of the horizontal canal is taken away. The portions of bone which cannot be filed off are removed by means of narrow gravers and scrapers (p. 4), and with these instruments the inner crus of the horizontal canal is likewise opened.

The preparation of the osseous capsule of the cochlea is now proceeded with. For this purpose the diploëtic bone tissue, extending from the point of the pyramid to the compact capsule, must first of all be removed, partly with forceps, partly with file and graver. The compact lamella of the posterior wall of the pyramid, however, which serves as a basis for the cochlea, and helps to form the internal auditory meatus, must be preserved intact, which is simply done by removing, with a broad graver, the remaining diploëtic osseous tissue between it and the compact capsule. The form of the cochlea, with its convolutions, can now be worked out, by means of gravers and files, from the exceedingly hard mass of bone surrounding the cochlear cavity, but as the boundary between the capsule and overlying bone in the adult can no longer be detected, all such preparations are to be regarded as artificial. In order, therefore, to dissect out the cochlear convolutions in their true positions, it is necessary to open them. Commencing at the fenestra rotunda, the lower convolution is opened with file and graver to a width of ½ mm., and from here the opening of the convolutions continued up to the proximity of the apex of the cochlea. Not until they are opened in such a manner that the edges of the lamina spiral. ossea can be distinctly seen through the slit, may the osseous mass which lies between the opened convolutions, and is most strongly developed between the first and second turns, be carefully filed out, in order to give the preparation externally the form of the cochlea.

For the preparation of the osseous labyrinth the dental engine is also well adapted, which, provided with variously shaped drills and bits, renders it possible with a little practice to shorten the proceeding of laying bare the labyrinthine capsule.

Model collections of preparations of the labyrinth are those of the anatomist Ilg, of Prague, one of which is in the museum of that city; the other, formerly in the Josephinum, is now in the anatomical museum of Vienna.

Laying Bare of the Vestibule.—If the roof of the vestibule, situated between the frontal (superior) semicircular canal and the cochlea, be removed by a horizontal cut, having its concavity upwards, and passing above the fenestra ovalis, a view into the vestibular cavity is obtained. It will reveal, posteriorly, the following: The three ampullar openings and the two mouths of the semicircular canals, anteriorly, the entrance into the scala vestib. of the cochlea, on the external wall (Fig. 96, o), the fenestra ovalis, on the inferior wall, the commencement of the lamina spiralis ossea (sp), and the fissure caused by the maceration of the first piece of the spiral membrane (lamina spiralis secundaria). This fissure forms, in the macerated temporal bone, the well-known communication between the vestibule and the scala tymp. of the cochlea.

This, however, is by no means sufficient for the study of the details of the vestibule and cochlea. In order to obtain an exact idea of the situation of the recess. hemiellipt. et hemisphæric., as well as of the ampullar openings and the mouths of the semicircular canals, of the labyrinthine fenestræ, etc., sections must be made through the vestibule and cochlea in three directions, on several petrous bones. The cuts are made with fine fret-saws, either on dry temporal bones, or on such as have previously been lying for twenty-four hours in water.

A frontal cut at the posterior edge of the fenestra ovalis through the vestibule brings into view on the posterior wall of the latter, and at its transition into the upper and lower walls (Fig. 95) : the ampullar orifices of the superior (as), of the horizontal (ah), and of the posterior (s) semicircular canals, as well as the mouths of the horizontal (h) canal and of the canalis communis (co), formed by the union of the superior and posterior canals, grouped as shown in the accompanying illustration.

On the anterior vestibular aspect of the frontal section are visible the commencing portion of the lamina spir. ossea, the entrance into the scala vestibuli of the cochlea, and medially from it a part of the recess. hemisphær.

A sagittal cut in the longitudinal axis of the pyramid through the

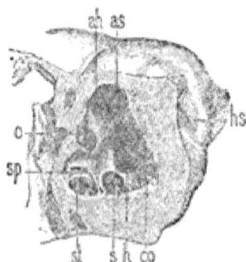

FIG. 95.—Frontal section through the vestibule : view of the posterior vestibular wall. Enlarged to double its size.—o = fenestra ovalis ; ah = ampulla horizontalis ; as = ampulla superior ; s = ampullary orifice of the posterior semicircular canal ; h = embouchure of the horizontal semi-circular canal ; co = common embou-chure of the frontal and sagittal canals ; sp = commencing portion of the lam. spiralis in the vestibule ; st = scala tymp. of the cochlea ; hs = hiatus subarcuatus. After a preparation in my collection.

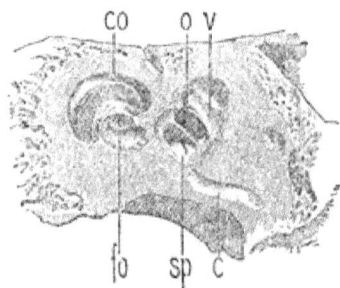

FIG. 96.—Section through the long axis of the pyramid : view of the lateral half. The cut passes through the middle of the vestibule and through the lower convolution of the cochlea. (Double its natural size.) — v = vestibulum ; o = fenestra ovalis ; sp = lower wall of the vestibule with the commencing portion of the lam. spiralis ; c = canalis semicirc. inf. ; co = lowest turn of the cochlea with the lam. spiralis ; fo = tractus spiralis foraminulentus. After a preparation in my collection.

middle of the vestibule, consequently parallel to the medial vestibular wall, shows, on the lateral section, the details of the external and inferior walls of the labyrinth (Fig. 96); the oval window (o) ; below this, on the floor of the vestibule, the before-mentioned fissure (sp) leading into the scala tymp., the ampullar openings of the superior and horizontal semicircular canals situated above the oval window on the outer wall of the labyrinth, the section of the lower crus of the posterior canal (c), and the oblique section of the lowest turn of the cochlea (co) with the tract. spiralis foraminulentus (fo).

On the medial wall of the vestibule (Fig. 97) are visible: parts of the recess. hemiellipt. et hemisphaeric., oblique sections of the semi-circular canals, the vestibular opening of the aquaeduct. vestibuli, and the sections of the meat. audit. int. and of the facial canal. As a matter of course, the view of the outer and the inner wall of the

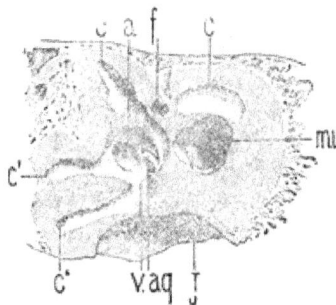

Fig. 97.—Sagittal section through the long axis of the pyramid: view of the inner half of the section.—v = inner wall of the vestibule with the recess. hemiellipticus; c c' c'' = segments of the section through the osseous semicircular canals; a = embouchure of the canalis communis; aq = vestibular opening of the aquaeductus vestibuli, with its furrow-shaped continuation downwards on the inner wall of the vestibule; mi = section of the meat. audit. intern.; c (front) = portion of the lower convolution of the cochlea; f = section of the canalis facialis; j = fossa jugularis.

labyrinth will vary, according to the section falling nearer to the medial or to the lateral wall, and according to the more or less slanting direction of the cut.

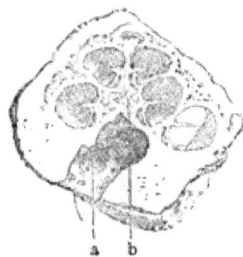

Fig. 98.—Section through the osseous capsule and the modiolus of the coch-lea, with the lamina spiral. ossea.—a = internal audi-tory canal; b = modiolus.

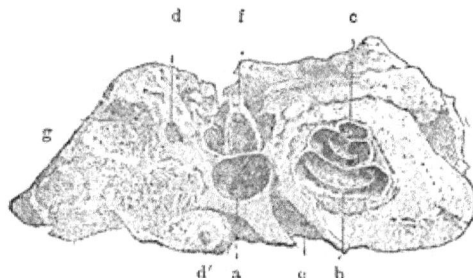

Fig. 99.—Horizontal section through the petrous bone of a new-born infant. (Enlarged to double its size.)—a = vestibule; b = base of the cochlea; c = cupola; d d' = sections of the superior semicircular canal; e = internal meatus; f = stapes; g = antrum mastoideum.

Sections through the cochlea are made either in a frontal direction, perpendicularly to the long axis of the pyramid, or horizontally, note being taken of the fact that the apex of the cochlea lies level with the canal. pro tens. tymp. (see Fig. 42, p. 40). The interior of the cochlea appears different, according to whether the scalae have been

opened by the cut laterally from the modiolus, or the latter divided
exactly in its axis.   This section (Figs. 98 and 99) gives a clear representation of the relationship of the modiolus to the lamin. spiral. ossea,
and of the diameter of the former, on which the canals for the bloodvessels and nerves can be distinguished by magnifying power. Sections
which meet the cochlea outside of the modiolus bring under observation the external surface of the latter and the relative inclination of
the osseous spiral lamina winding round it.   Instructive preparations
of this kind, showing also the relationship of the hamulus of the
lamina spiralis to the apex of the cochlea, are obtained if the cochlear
capsule, taken from the temporal bone of a child, be broken off in
small particles with a pair of small, finely-pointed forceps, special care
being taken to keep the lamina spiralis intact.

Topographical sections of the entire labyrinth, which demonstrate
the relative position of the various parts of the labyrinthine cavity, are
made in a horizontal or sagittal direction through the pyramid.   The
best horizontal sections through vestibule, cochlea, and internal
meatus are obtained if the cut lies in the middle of the latter, and a
little above the fenestra ovalis. Cuts which pass somewhat lower,
through the middle of the oval window, meet the cochlea only at its
lower turn.   Two parallel sagittal cuts through the labyrinth in the
longitudinal axis of the pyramid are sufficient to make us acquainted
with the above-described details in the vestibule, as also with those of
the cochlea, which, being divided perpendicularly to its axis, will show
the transverse section of the modiolus.

(b) *Preparation of the Membranous Labyrinth.*

The preparation of the membranous labyrinth is a most difficult
task ; but the obstacles which the first attempts meet with are soon
overcome with practice.   It was Voltolini (*l.c.*) who first pointed out a
method by which the three membranous semicircular canals, in connection with the utriculus, can be removed entire from the labyrinthine cavity. The proceeding, with some modifications of Voltolini's
method, is as follows : The canal. semicirc. superior is first opened by
carefully removing, with a moderately strong file or short scraper, the
osseous mass on the eminentia arcuata, until the canal shines through
the thin layer of bone.   In the same manner the bone covering the
external and internal crura of the superior, as well as the entire
posterior canal, is cleared away by careful filing and scraping, until
both canals can be seen through.   Next, the bone lying in front of
the superior canal and over the vestibule is removed in thin layers,
at first with the chisel, but in advancing deeper, with the short

scraping-iron, until the vestibular cavity also shines through. Now, at the spot where the superior and posterior canals, joining in a common opening, enter the vestibule, a small hole about the size of a pin's head is made in the roof of the vestibule with the point of a small knife. Through this a thin preparing needle is introduced, with which the bridges of connective tissue, uniting the utricle and the ampullæ with the osseous walls of the vestibule, are loosened. In order not to injure, during this manipulation, the saccules and ampullæ, the point of the needle should be kept close to the osseous wall, and the loosening of the connective-tissue bridges effected with care by slight lateral movements. Not until then can the roof of the vestibule be carefully broken open from this aperture, with a short pointed knife.

After this the superior and posterior canals, already shining through, are opened by very cautiously cutting with a knife the thin layer of bone still remaining, avoiding, in doing so, any injury to the membranous canals. The whole preparation is now placed under water in a wide saucer, and with a preparing needle, somewhat bent at its point, the membranous canals, which are only slightly adherent to a circumscribed spot on the bony wall (see the Histological part), are cautiously lifted out of the osseous canals, without separating their connection with the vestibular structures. Only when both membranous canals are freely floating in the water is the preparation again removed from it, the two canals being carefully pushed into the vestibule with a needle.

To prepare out also the horizontal membranous canal, in connection with the utriculus, the now superfluous superior bony canal and the upper crus of the posterior one, together with the osseous mass between them, must be cleared away by a horizontal cut. To avoid injury to the horizontal canal, this cut ought to be carried a little above the outer crus of the latter, which projects behind the fenestra ovalis on the inner wall of the tympanic cavity. Slow scraping and filing of the surface of the cut will soon render the bone so far transparent as to allow a view of the horizontal canal, which, under the already described precautions, is opened out up to the vestibule, whereupon the membranous canal is lifted out under water, and in its turn pushed into the vestibule.

Semicircular canals, ampullæ and utriculus, now form one lump in the vestibule. In order to remove intact the whole mass, the adhesions still existing between the membranous structures and the osseous vestibular wall, and which can be perceived under water with the aid of a lens, are carefully loosened with a needle, keeping the point close to the osseous wall, and then the entire lump is allowed to glide out of the vestibule into a watch-glass filled with water. Here the labyrinth is disentangled, not with needles, but by gently shaking the watch-glass and by allowing water to fall in drops upon it from a certain height, until saccules and membranous canals have assumed their normal position. The preparation may either be preserved in spirits of wine in a correspondingly small glass jar (Fig. VI. of the plate), or mounted as a miscroscopic object. This method is recommended only for the purely anatomical representation of the membranous labyrinth. For pathological cases, however, where it is not only a question of observing the finer structural changes of the membranous labyrinth, but also of ascertaining the presence of pathological processes on its osseous wall, and the relation of this to the membranous labyrinth (new formations of connective tissue and of bone in the labyrinth), the histological examination on decalcified sections of the labyrinth is to be preferred. (Compare the Histological part.)

If, in the examination of the membranous labyrinth, no regard is to be paid to the connection of the membranous structures, the semicircular canals and ampullæ may be obtained, though unconnected with the utricle, in the following simple manner : on the pyramid of a fresh temporal bone, placed upon a firm support, a chisel, 2-3 cm. broad, is applied a few millimetres in front of the eminentia arcuata superior, that is, above the middle of the vestibule in a frontal direction, and with one blow of the mallet the entire labyrinth is cut in two. The two pieces are placed in water, and waved to and fro in it, until portions of the utricle, of the ampullæ, and the ends of some of the membranous canals, are seen floating in the water. Although the last-named structures are fixed to the wall of the osseous canal at a circumscribed spot, they are very easily detached. If the ampulla be now taken hold of with a fine pincette, it is possible by careful, not too sudden pulling, to draw the three membranous canals from the osseous, and to subject them to microscopic examination. Of the utricle and saccule, it is true, only fragments are obtained, which, however, in some cases, are sufficient for pathologico-histological examination.

But should it be intended to obtain the saccules as intact as possible, irrespective of their connections with the semicircular canals, the attachments of connective tissue between the saccules and the walls of the labyrinth are loosened with a fine needle, through a small opening made in the upper labyrinthine wall after the manner described at the commencement, and the aperture having been widened, the still remaining connections are separated, and the saccules, under water, taken out of the vestibular cavity. The utricle may thus frequently be obtained quite intact, the saccule, however,

being more intimately connected with the labyrinthine wall, only in
fragments. The removal of the saccules from the vestibule by way
of the oval window, when the stapes and surrounding osseous mass
have been taken away, is only successfully accomplished after long
practice and by careful preparation, notwithstanding the fact that
between the external wall of the labyrinth and the saccules there
exists a considerable perilymphatic space (Steinbrügge), which is
favourable for laying them bare from the lateral side.

Rüdinger (*l.c.*) advises, for the purpose of general study of the fresh
labyrinth in the dissecting-room, applying a moderately-sized chisel
horizontally to the lateral edge of the internal meatus at its upper
part, and driving it in horizontally from behind forwards. The upper
half of the osseous labyrinth is thus forced away, and the vestibule
and cochlea are opened. Fragments of the membranous canals, of
the saccules and of the cochlea, can then be taken out under water
with needle and pincette.

The preparation of the cochlea in the fresh state presents difficulties in so far as the opening of the capsule is necessarily attended
with injury to its membranes. For the study of the structure of the
cochlea, however, its anatomical preparation is indispensable. In
order to obtain at least part of the lamina spiralis membranacea
intact, after removal of the upper wall of the meat. audit. intern.
above the entrance of the ramus cochleæ into the modiolus, the
lowest convolution of the cochlea is opened with a narrow chisel or
graver. The opening is enlarged far enough to allow a slightly
bent needle to be introduced, with which the broader margin of the
membranous spiral lamina is detached from the outer cochlear wall.
This having been done at the superior segment of the lowest turn of
the cochlea, that portion of the capsule is broken away in pieces with
a small, finely-pointed bone forceps, and from here the loosening of the
lamina spiralis, the severing of the structural supports, and the breaking
away of the capsule in the remaining turns should be continued. Thus
the upper portion of the cochlea is exposed, while the lower part of
the convolutions still remains in connection with the capsule. An
attempt is now made with the preparing needle to loosen from above,
as far as possible, the lower portion of the lamina spiralis from the
bone also; then, with a small knife, the base of the modiolus is cut
through close to the internal meatus, after which the entire contents
of the cochlea, with the exception of the part of the first convolution bordering on the fenestra rotunda, can be lifted out with a fine
pincette.

If the cochlea, thus prepared out, be placed in dilute spirits of
wine, and the liquid gently shaken, the lam. spiral. membran. will

expand, and under moderate magnifying power a clear representation
is obtained of the osseous and membranous spiral plate in its connection with the modiolus, and portions of it may be cut off with scissors
to be examined microscopically.   For pathologico-anatomical research
this method is only applicable in exceptional cases, as for the study of
the histological changes microscopical sections of the entire cochlea
(see the Histological part) are infinitely better suited.

It is very easy to obtain the membranous cochlea, with its capsule
complete, from fœtal organs (five or six months), by splitting with two
preparing needles the still soft cartilaginous capsule, and detaching it
from the membranous part.   It is also possible, on the petrous bones
of new-born infants, less frequently on those of adults, to prepare out
intact the membranous cochlea from its osseous capsule, if the pyramid
be first hardened by placing it for two or three weeks in a 2% solution
of chromic acid, and then slowly decalcified with nitric or hydrochloric
acid to such a degree that the capsule can be removed in layers with
needle and pincette.   My collection contains several preparations of
the cochlea obtained in this manner.

Recently Katz* has published a method of representing the membranous labyrinth by which, through corrosion of its osseous capsule,
the membranous structures are exposed.   A fresh petrous bone, or
even an old spirit preparation, is decalcified for six to eight days—if
from an adult in $15\text{-}25\%$, if from a child in $8\text{-}12\%$, hydrochloric acid ;
then $10\text{-}15\%$ nitric acid is added to the liquid, and the preparation
allowed to remain in it for eight to fourteen days longer, until it
appears changed into a gelatinous mass.

The corrosive liquid having been poured off, the preparation is
placed in water, where the softened bone is carefully removed with
needle and camel's-hair brush, so far as to leave only enough of
it to hold the preparation together.   By this means are obtained, if
not always entirely, at least partially in connection, the membranous
saccules, ampullæ, and semicircular canals ; also, especially clear, the
twigs of the ramus vestibuli, running in the petrous bone ; and, lastly,
the facial nerve.   For preserving the preparation a 2% solution of
chromic acid is recommended.   In the experiments made by me, I
always obtained mere fragments of the membranous labyrinth.

Another method, recommended by Barth, for representing the
membranous labyrinth by corrosion, is to be found in the chapter
' The Preparation of the Organ of Hearing by Corrosion.'

Very instructive for the macroscopical study, and for the topography
of the membranous labyrinth, are those sections of decalcified petrous
bones on which the labyrinthine structures have been hardened before

* Beitrag zur anatomischen Präparation des häutigen Labyrinths (M. f. O., 1887, Nr. 7).

decalcification; while, as regards the chemico-technical proceeding, we refer to the Histological part of this work. We here only observe that, when the decalcified pyramid, dehydrated by alcohol, is placed in a solution of celloidin, the latter penetrates into the labyrinthine cavity; and after the hardening of the medium in diluted alcohol (70 %), the membranous structures of the labyrinth are so fixed as to appear in their original positions on the sections.

Especially suitable for such sections are petrous bones of new-born infants, which are decalcified in a much shorter time than those of adults. Horizontal sections through the internal meatus, modiolus, vestibule, fenestra ovalis, and stapes yield very beautiful preparations,

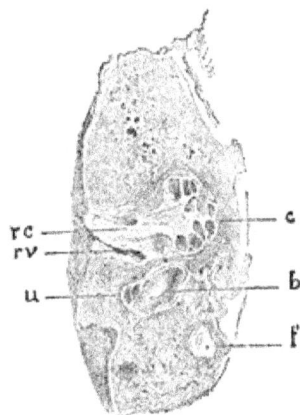

FIG. 100.—Horizontal section through a hardened, and afterwards decalcified petrous bone of a new-born infant : lower half.—u = transverse section of the lower portion of the utricle; b = floor of the vestibule, with the commencing portion of the lamina spiralis ; c = cochlea ; rc = ramus cochleæ ; rv = ramus vestibuli.

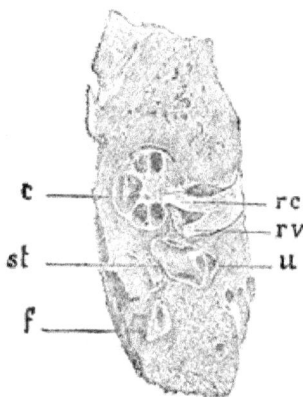

FIG. 101.—Upper half of the same preparation. — u = the utricle, situated in the vestibule ; st = transverse section of the foot-plate of the stapes ; c = cochlea ; rc = ramus cochleæ ; rv = ramus vestibuli. After a preparation in my collection.

which, preserved in dilute spirit, are of much service for demonstration purposes, especially when magnified with a lens.

The accompanying illustrations (Figs. 100 and 101) give a clear idea of the relationship of the membranous structures of the labyrinth. On the upper segment of the pyramid (Fig. 101, u) we notice, through the transparent celloidin, the transverse section of the utriculus (u) filling, in the shape of an elongated sac, the greater portion of the vestibule, while in the lower segment (Fig. 100, u) it occupies only a small part of the medial space of the latter. The section through the cochlea, modiolus (c), and the internal meatus, allows the arrangement of the cochlear convolutions, the mode of attachment of the membranous spiral plate, and of Reissner's membrane to be recognised;

finally, in the internal meatus the nerv. acusticus, with its division
into the ramus vestibuli (rv), and ramus cochleæ (rc), is visible.
The nerve bundles, sharply marked by their light-green colour,
may, with a magnifying lens, be traced on the one side into the
vestibule, and on the other, through the modiolus, into the lamina
spiral. ossea.

For the study of the topographical position of the semicircular
canals, ampullæ and saccules, it is further advisable to make a series
of frontal sections through the previously decalcified pyramid, and
impregnated with celloidin after the method described. By carrying
a frontal cut through the posterior portion of the pyramid, 2-3 mm.
behind the superior semicircular canal, a preparation is obtained, on
which will be seen, as the accompanying illustration (Fig. 102) shows,

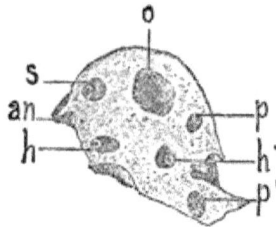

Fig. 102.—View of the topographical position of the membranous semicircular canals within
the osseous : transverse section through the posterior part of the decalcified pyramid of
a new-born infant. (Right ear.)—s = transverse section of the superior semicircular canal;
h h' = transverse sections of the horizontal semicircular canal ; p p' = transverse sections
of the posterior semicircular canal ; o = fossa subarcuata ; an = inner wall of the mastoid
antrum.   After a preparation in my collection.

the relation of the diameter of the membranous to the osseous canal,
and the mode of attachment of the membranous canals to the osseous
wall. A frontal cut carried more anteriorly, close behind the superior
canal and behind the fenestra ovalis (see Fig. 103), meets the posterior
portion of the vestibule, in which the section of the utriculus (u) with
the cisterna perilymphatica, placed between it and the external
vestibular wall, are perceived ; likewise the ampulla of the posterior
canal (ai), that of the horizontal canal (ah), the inner crus of the
latter (ch), and a part of the intratemporal aquæduct. vestibuli (aq).

12. PREPARATION OF THE AQUÆDUCTUS VESTIBULI ET COCHLEÆ.
(Ductus endo- et perilymphaticus.)

The representation of the aquæducts of the labyrinth is one of
the most difficult tasks in the preparation of the organ of hearing.
This may be inferred from the remarkable fact that the important
anatomical discovery of the highly-gifted Domenico Cotugno, and his

minute description of the labyrinthine aquæducts,* illustrated by diagrams, remained for nearly a whole century almost unnoticed by anatomists.   In spite of the corroborating statements of F. Meckel, Hyrtl,† and Van den Broeck, in whose excellent, though little known anatomical atlas of the organ of hearing,‡ Plate IX., the intradural sac of the aquæduct. vestib. is beautifully delineated, the vestibular aquæducts were, until a few decades ago, looked upon as venous canals, or osseous fissures containing processes of the periosteum. Cotugno's discovery was fully confirmed by the excellent monograph,

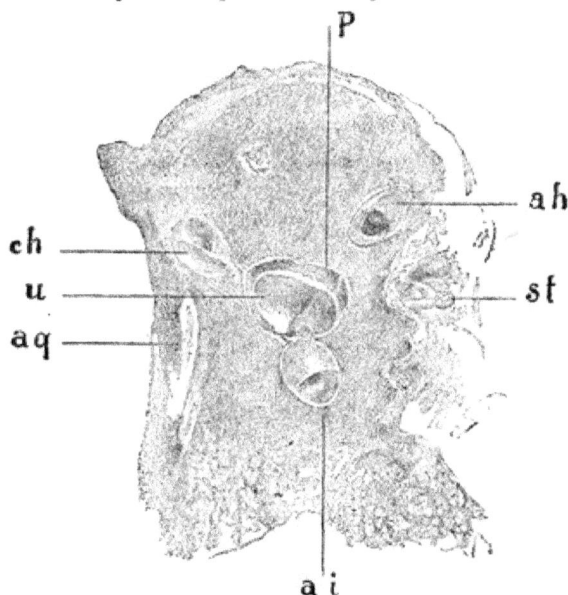

Fig. 103.—Frontal section through the posterior portion of the vestibule, behind the fenestra ovalis.—u = utriculus ; p = cisterna perilymphatica ; ah = ampulla horizontal. ; ch = section of the horizontal semicircular canal ; ai = ampulla inferior ; st = stapes ; aq = section of the aquæductus vestibuli.   After a preparation in my collection.

' Entwicklung und Bau des Gehörlabyrinths' (Leipzig, 1863), of A. Böttcher, to whom we are indebted for having discovered the membranous canals of the aquæduct. vestibuli discharging into the saccules; further, through the anatomical studies of C. Hasse (1870-73), through the works of E. Zuckerkandl and Weber-Liel (M. f. O., 1869, Nr. 8), and by Rüdinger's§ more recent publication; and it is to Zuckerkandl

* De aquæductibus auris humanæ internæ.   Viennæ, 1774.
† Vorläufige Mittheilungen über das Knochenlabyrinth der Säugethiere.   Med. Jahrbücher, 1843.
‡ Ontleekundige en Physiologische Beschrijving van het zintuig des Gehoors door Dr. J. K. van den Broeck (Arnhem, 1853).   A copy of this rare work was presented to me by Dr. Moll, of Arnhem.
§ 'Ueber die Abflusscanäle der Endolymphe des inneren Ohres ' (Sitzungsbericht der kgl. bayer. Akademie der Wissensch., 1887, Heft 3).

in particular that credit is due for having, by his excellent treatment of the subject, supplied definite data for the preparation of the aquæductus vestibuli.

(a) Aquæductus Vestibuli.

If a cut be carried by means of a saw through the pyramid of the petrous bone, corresponding to its long axis, and parallel to its posterior surface, thus dividing the vestibule into halves, we find, on the inner vestibular wall, in front of and below the common embouchure of the superior and posterior semicircular canals, the small internal aperture of the aquæductus vestibuli (Fig. 104), terminating below in a furrow. This opening leads into a very narrow canal, which is rarely passable to a bristle, and which describes a slight curve in the osseous mass of the petrous bone lying in front

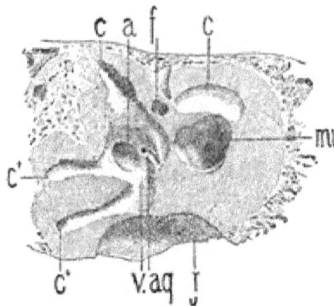

Fig. 104.—Sagittal section through the long axis of the pyramid : view of the inner surface of the cut.—v = inner wall of the vestibule, with the recess. hemiellipticus ; aq = vestibular embouchure of the aquæduct. vestibuli, with its furrow-shaped continuation downwards on the inner vestibular wall ; f = section of the canalis facialis ; j = fossa jugularis.

of the posterior (sagittal) semicircular canal, and terminates at the posterior surface of the pyramid behind the porus acustic. intern. in a slit-like dilatation (apertura ext. aqueduct. vestibuli) (Fig. 21, p. 24).

This osseous canal contains the extremely narrow membranous canaliculus of the aquæduct. vestibuli. To expose the former in the macerated temporal bone, from the apertura extern. aquæduct. vestibul. (Fig. 105, av) to its opening into the vestibule (o), the osseous lamella, bordering posteriorly on its external opening, is removed by means of a broad graver, and the triangular space of this slit (av) laid bare, up to the narrow part of the osseous canal. From this point the work of preparation must be limited to the removal of small pieces only, by means of a pointed file and fine graver, because the narrow canaliculus is so easily choked with bone dust, that the trace of it is lost in the mass of the petrous bone, and difficult to find again.

In order, therefore, not to miss the direction of the course of the small canal during the preparation, it is advisable first to inject it with coloured wax. This is best done by inserting a small tube air-tight into the fenestra ovalis, the round window being at the same time closed up with putty. The injection may be considered complete when the coloured wax can be seen at the hiatus aquæduct. vestibul. on the posterior wall of the pyramid. When this is not the case, which happens rather frequently, the cause of the failure must be sought for in the blocking up of the canal by tissue-remains or other products of maceration.

The colouring of the osseous canaliculus, for the purpose of more easily finding it during the preparation, succeeds on well-macerated

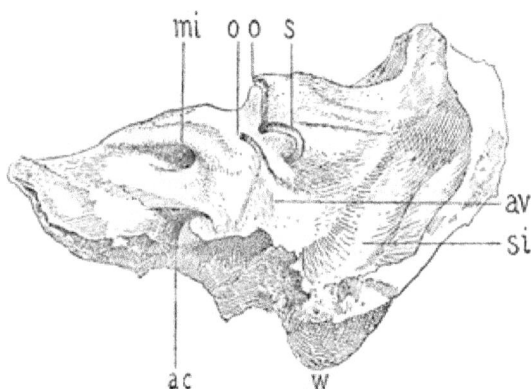

Fig. 105.—Posterior surface of the pyramid of the right temporal bone, with the opened intratemporal canaliculus of the aquæduct. vestibuli and the embouchure of the aquæduct. cochleæ.—mi = meat. audit. intern. ; av = slit-like opening of the aquæduct. vestib. on the posterior surface of the pyramid ; o = opening of the osseous canal of the aquæduct. vestib. in the vestibule ; o' = canal. semicirc. super., opened ; s = canal. semicirc. post., opened ; ac = funnel-shaped outlet of the aquæduct. cochl. ; si = sinus transv. ; w = proc. mast. After a preparation in my collection.

temporal bones by placing the object in a coloured liquid, after previously thoroughly soaking it in water for one or two days ; the liquid will penetrate into the fine osseous canals without colouring the bone itself. Carmine or hæmatoxylin solutions are best adapted for the purpose ; the colour of the latter can be eliminated, when the preparation is completed, by placing the bone in dilute nitric acid.

On temporal bones of new-born infants, the two osseous aquæducts may be successfully prepared out from the surrounding bone, with graver and file. Several perfect specimens of this kind are to be found in the celebrated collection of Professor Ilg, which has been added to the anatomical museum of Vienna. The accompanying illustration (twice the actual size) is taken from a preparation in this collection.

The preparation of the membranous aquæductus vestibuli (ductus et saccus endolymphaticus), is divided into that of the intracranial, and that of the intratemporal portion. The tracing of the intracranial portion, a sac-shaped space, 5-9 mm. broad and 8-15 mm. long, situated between the layers of the dura mater covering the posterior surface of the pyramid, presents, after a little practice, no special difficulty. A most important guide is the protuberance, 3-4 mm. behind the porus acusticus internus, which is found in front of the apert. ext. aquæduct. vestibuli, and which, as in the macerated bone, also projects in the posterior surface of the pyramid covered by the dura mater. From this protuberance, in a slanting direction backwards and a little downwards towards the lower curvature of the sinus lateralis, lies the intradural sac of the vestibular aquæduct alluded to. In some preparations, as I repeatedly noticed, its position

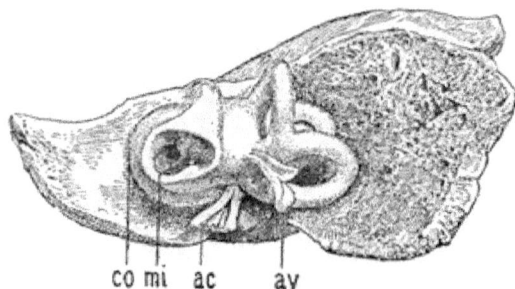

FIG. 106.—Posterior view of the osseous labyrinth, with the three semicircular canals, the cochlea and the aquæducts of a new-born infant. (Enlarged to double its size.)—co = cochlea ; mi = meat. audit. intern. ; ac = aquæductus cochleæ ; av = aquæductus vestibuli.

and extent can be clearly distinguished by a slight fluctuation, which becomes perceptible when the place is touched with the point of a probe. In order to expose the sac, a short incision is made behind the protuberance, i.e., at the point of transition of the sac into the intratemporal portion, and air is forced into it through a small tube introduced into the opening. If this is successful, the inflated sac will show itself sharply defined in its limits and extent.

With the object of laying bare the inner surface of the sac, a whalebone sound is now inserted into the opening of the incision, then carefully moved backwards and downwards, and the medial wall of the sac is cut in its longitudinal direction, by means of fine scissors. For the purpose of exposing the floor of the sac (lateral wall), the slit in the medial wall is divided by two transverse cuts, and the four triangular flaps thus made are turned back, as shown in Fig. 107, av.

In order to demonstrate the capacity of the sac, Zuckerkandl suggests making an incision and stuffing the whole cavity with cotton wool, whereby the entire intracranial portion of the aquæduct in its outlines stands out in relief on the posterior surface of the pyramid. The sac may also be injected with a quickly setting coloured mixture of wax and resin, if an incision be made at its most anterior boundary, to receive the fine canula of a syringe.

In the above-mentioned treatise of Cotugno there are figured several lateral canals branching off from the intradural sac and running on in the dura mater. The existence of these small canals, passing off from the floor of the sac, was confirmed by Zuckerkandl and Weber-Liel. Recently Rüdinger (l.c.) has proved, by a series of microscopical sections on embryonic heads of various animals, that from the floor of the endo-lymph sac tubular prolongations pass out

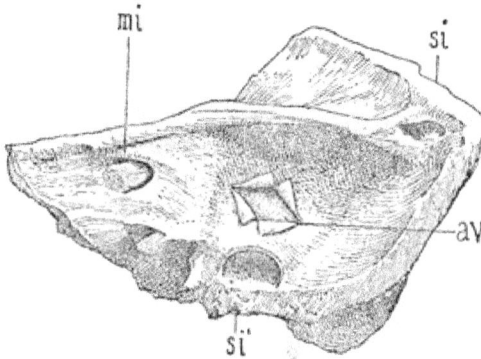

FIG. 107.—Posterior surface of the pyramid of the right petrous bone of an adult, with the opened intradural sac of the aquæduct. vestibuli.—mi = meat. audit. internus, with the nerv. acusticus ; av = opened intradural sac of the aquæduct. vestibuli; si = upper section of the sinus transv. ; si' = its lower section. After a preparation in my collection.

and penetrate the dura mater, and he regards these as channels for the discharge of the endo-lymph from the membranous labyrinth. These tubes are, according to Rüdinger, completely surrounded by large and wide lymph sheaths, to which externally the connecting substance of the dura mater is closely attached.

Considerably more difficult is the representation of the intra-temporal membranous tube of the aquæduct. vestibuli, which runs in the petrous bone. Even after having repeatedly prepared it, one frequently fails to discover this small tube, which, in the form of a slender connective-tissue cord, passes through the densest portion of the petrous bone and can only be demonstrated in its continuity by carefully removing the bone in thin layers with small sharp chisels and scraping irons. The work is materially facilitated if the object has previously been lying for several weeks in a 10% solution of chromic acid, when the connective tissue enclosing the aquæduct contrasts, by its yellow colour, with the surrounding bone. The position and direction of the small canal within the bone can also be made perceptible by injecting it with a coloured liquid, provided the intradural sac of the aquæduct can be successfully opened at its posterior part, so that after loosening the dura mater from the subjacent bone, the nozzle of a canula may be tied in the opening, and the injection afterwards proceeded with. A still more simple method of making the small membranous tube traceable in the bone by colouring is, according to Weber-Liel, to dip the preparation, with the opened intradural sac of the aquæduct, into a coloured liquid, and through a small opening in the superior semicircular canal to aspirate the labyrinthine fluid by means of a small glass tube until the coloured liquid (Beale's blue) appears in it.

Still easier is the preparation of the aquæduct. vestibuli on petrous bones which have been previously placed for two to three weeks in dilute chromic acid, and afterwards decalcified (see decalcifying fluids in the Histological part). By removing the decalcified bone in layers with a flat scalpel or razor, it is possible in a short time to clear the entire intratemporal portion up to its discharge into the vestibule. For safety, it is advisable to begin the preparation at the hiat. aquæduct. vestibuli, and to continue from here along the easily-discernible strand of connective tissue, the removal in layers of the petrous mass as far as the vestibule. In such preparations, if preserved in spirit, the light-green connective-tissue strand of the aquæduct is sharply defined from the dark-green colour of the bone.

The representation of the commencing canaliculi of the aquæduct. vestibuli discharging into the saccules, and which Böttcher discovered by means of serial microscopical sections, is so difficult, that Zucker-kandl, after many attempts, only succeeded once in seeing them distinctly. By further improvements in the technical details of the methods of injecting, the difficulties of anatomical demonstration may be removed.

(b) Aquæductus Cochleæ (Ductus perilymphaticus).

The laying bare of the osseous aquæduct. cochleæ, which commences with a small opening behind the groove of the fenestra rotunda on the lower wall of the first cochlear convolution, and terminates, at the edge between the posterior and inferior surfaces of the pyramid, with a funnel-shaped dilatation (Fig. 105 av), is even more difficult than that of the aquæduct. vestibuli.

For the preparation of the small canal on the macerated temporal

bone, it is here also advisable, according to the method already described, to colour its walls by injecting with, or by dipping in a staining
fluid.   We begin by chiselling away the bone at the above-mentioned
funnel-shaped outlet of the aquæduct at the inferior edge of the
pyramid.   The exposure of the lower and wider portion of the canal
is quickly accomplished; much more difficult is it to trace the upper,
very narrow, straight, or somewhat curved piece up to its entrance
into the scala tymp., where the work must be done mainly with the
graver and finely-pointed files.

The method of representing the aquæduct. cochleæ on unmacerated
temporal bones is the same as that of the preparation of the aquæduct. vestibuli.   The colouring of the narrow tube is here also effected
by dipping the funnel-shaped outlet in coloured liquid, and by
aspiration from the superior semicircular canal, but care should be
taken to avoid injuring the membranous canal while opening the
osseous one.   According to Weber-Liel it is also possible, by aspiration
of the air in the external meatus, to fill the scala tymp. from the
aquæduct. cochleæ; frequently, however, this experiment proves a
failure, like the filling of the aquæduct. cochleæ by injection of
the sub-arachnoideal space, with which the perilymphatic spaces
of the labyrinth communicate.

Most expeditiously will it be accomplished on decalcified petrous
bones, if the bone be removed in layers with a scalpel.   It is best to
commence at the lower funnel-shaped outlet of the aquæduct, where
the triangular, at the narrow portion tapering, fasciculus indicates the
direction to be followed in removing the layers of the softened bone,
up to the embouchure of the canal into the lower scala of the cochlea.

The representation of the aquæducts by means of corrosion will
be discussed in the section 'Making Preparations of the Organ of
Hearing by Corrosion.'

13. PREPARATION OF THE BLOODVESSELS AND NERVES OF THE ORGAN OF HEARING.

PREPARATION OF THE BLOODVESSELS AND LYMPHATICS.

Most suitable for the preparation of the bloodvessels of the organ
of hearing are those objects which have been injected, and in which
the injection substance, partly by its glimmering through, and partly
by its filling the tube of the vessel, materially facilitates the detection
of even the finer branches.   The injection is always made on perfect
skulls, in adults from the carotis communis, in children preferably
from the aorta.

Instead of the hitherto generally used substance made of wax, with the addition of turpentine, tallow, and olive-oil, the anatomical institutions now employ one of putty, introduced by Teichmann, with which the preparations are injected in the cold state. According to Teichmann, not only fresh objects, but also older ones, or such as have been kept in spirit, are suitable for injection. The composition of this easily-prepared mass being generally little known, the proportions, as given by Teichmann, may here be noted.

Red Mass for injecting the Arteries:

Whiting	.	.	.	. 5.00 g.
Vermilion	.	.	.	. 1.00 g.
Inspissated linseed-oil.			.	0.9-1.00 ccm.

This well-triturated mass, large lumps of which may be kept in water, is, before being used, dissolved in 0.75 ccm. of carbon bisulphide, and this solution slowly injected by means of special screw syringes. Should a large quantity of the substance be required, the proportions of the various ingredients, as given above, should be strictly adhered to. Only in those cases where the capillary vessels also are to be injected is the finely-ground mass, free from grit, to be thinned by a double quantity of carbon bisulphide. Capillary injections of the membrana tympani, of the tympanic mucous membrane, and of the labyrinth, it is safest to make by means of a coloured glue substance (Berlin blue, or carmine). If it be intended to inject also the veins and venous sinuses in the temporal bone, Teichmann's blue composition, made up as follows, may be used: Oxide of zinc 15.00 g., ultramarine 1.00 g., boiled linseed-oil 2.0-2.5 ccm., carbon bisulphide or sulphuric ether 1.00 ccm. To inject the larger lymphatic vessels, Teichmann recommends a yellow substance consisting of: Oxide of zinc 2.00 g., inspissated linseed-oil 3.0 ccm., and sulphuric ether 2.0 ccm.

PREPARATION OF THE ARTERIES OF THE ORGAN OF HEARING.

The preparation of the arteries of the external auricular region and of the auricle is best made with that of the arteria carotis extern. and its branches. After exposing, on a well-injected preparation, the upper part of the arteria carotis extern. and finding, below the condyloid process of the inferior maxilla, the place where it divides into the arteria maxillaris intern. and arteria temporalis superficialis, the last-named branch, which ascends perpendicularly in front of the tragus in a nearly vertical direction across the root of the zygomatic arch, is dissected out, after which the laying bare of the rami auriculares anteriores, passing from the external or posterior periphery of the arteria temporalis superficialis towards the auricle, is easily effected. Two or three inferior branches of the arteriæ auriculares anteriores

supply the lobe, the tragus, and the anterior cartilaginous wall of the
meatus; several superior branches, the upper part of the helix.
Somewhat more difficult is the preparation of the arteria auricularis
posterior, arising from that portion of the arteria carotis extern.,
which passes through the parotid gland. To find this arterial branch
at its point of origin, the carotis extern. should first be exposed by
dissecting away the parotid tissue and the arteria auricular. post.,
arising behind the posterior belly of the digastric muscle, must be
carefully prepared out. The preparation of the superior branches of
the arteria auricularis post., ramifying on the posterior surface of the
auricle, presents no difficulty.

Small twigs of the arteria auricularis post. penetrate between the
antitragus and the processus helicis caudatus, and through the
vascular canals of the auricular cartilage, also to the anterior surface
of the auricle, where the finer ramifications spread towards the superior
and lower portions.

By injecting the finer vessels it is also possible to trace the arteria
stylo-mastoid., which is given off from the arteria auricular. post.,
and which enters the foramen stylo-mastoid., supplies the facial nerve,
and with a small branch passes through the canaliculus chordæ into
the tympanic cavity, anastomosing here with the vessels of the tympanic membrane.

The arteries of the external meatus, which in the outer portion of
the cartilaginous part arise from the arteriæ auriculares anteriores et
posteriores, but in the deeper portion of the cartilaginous and osseous
canal from the arteria auricularis profunda, can only be represented
on such preparations where the injection fluid has penetrated into the
finest arterial branches. As the bloodvessels mainly run along the
posterior superior wall, it is necessary, in order to expose them, to
remove, on an injected preparation hardened in alcohol, the lower wall
of the cartilaginous and osseous meatus by a horizontal cut carried as
far as the neighbourhood of the tympanic membrane, and to take off
the thick epidermic layer, covering the bloodvessels with a camel's-
hair brush. If the injected fluid has also penetrated into the
radiating vascular ramifications of the membrana tympani, it will be
noticed that a strong arterial twig, running along the upper wall of
the meatus, passes at the posterior boundary of Shrapnell's membrane
on to the tympanic membrane, where it can be traced along the
posterior margin of the manubrium down to the lower extremity.

PREPARATION OF THE ARTERIA AURICULARIS PROFUNDA.

In order to find this artery, the masseter muscle, together with the
zygomatic arch sawn through before and behind, must be turned down

towards the angle of the maxilla, the terminal division of the arteria carot. ext. having already been exposed by the previous preparation. By sawing the ramus of the lower jaw transversely through, about midway between the incis. semilunar. and the maxillary angle, and dividing the neck of the maxilla below the capitul., a piece of bone is made free, which is to be turned over in an upward direction, after the tendon of the temporal muscle has been completely cut through, whereby the commencing part of the arteria maxillaris intern. will become exposed. At the posterior periphery of the last-mentioned artery arises the arteria auricularis profunda, the branches of which can be traced, by careful preparation, to the deeper parts of the external meatus.

The arteria tympanica, also arising from the arteria maxillaris intern., can only be exposed up to its entrance into the Glaserian fissure, after disarticulation of the condyle of the inferior maxilla.

PREPARATION OF THE ARTERIA AUDITIVA INTERNA.

This takes its origin mostly from the arteria basilaris, sometimes, however, from the arteria cerebelli anter. Where it arises from the arteria basilaris, it crosses the medulla oblongata longitudinally. The origin of the artery may be best seen on a brain which has been injected *in situ* and carefully taken out of the cranial cavity, the artery being cut through close to the porus acust. int. along with the nerve trunks entering the latter. If, however, it be desired to retain it in its continuity, in order to follow it to the floor of the internal meatus, it is advisable, after taking away the roof of the skull, the cerebral hemispheres, and the entire tentorium, to remove the cerebellum, pons and medulla oblongata in pieces, in such a manner that the basilar arteries and their ramifications be retained on the base of the skull. The arteria auditiva int. is then seen passing from its origin at the base of the skull in a lateral direction to the meat. audit. intern. With the object of examining the further course of the artery towards the fundus, the superior wall of the internal meatus is removed with chisel and forceps. For details we refer to the section 'Preparation of the Acoustic and Facial Nerves.'

PREPARATION OF THE VEINS OF THE ORGAN OF HEARING.

It is much more difficult to completely inject the veins of the auditory apparatus than it is in the case of the arteries. It is not sufficient, as in the filling of the arterial system, to inject from one single spot, but it should be done from several points. As the venæ auditivæ int., which accompany the artery of the same name, the veins of the labyrinth and of the tympanic cavity enter, either directly or through the intervention of the meningeal veins, into the neighbouring sinuses of the dura mater (sinus petro-squamosus, petrosus superior, petro-basilaris and transversus), they are best injected from these sinuses. Most suitable for this purpose is a head severed from a very lean and anæmic subject, on which one or both venæ jugulares communes are injected with Teichmann's blue or yellow mass. In the latter case, for the purpose of injecting both jugular veins simultaneously, a bifurcated canula is made use of. Before the injection, the vertebral canal and the foramina transversaria are firmly plugged with cotton wool, and the stumps of the veins, appearing at the surface of amputation, are ligatured. Should, during the operation, the injection fluid ooze out at any part on the amputated surface, the open vessel must be twisted or ligatured.

The veins of the external ear, injected after this method, are, as a rule, more or less incompletely filled, as the valves prevent the advance of the medium towards the periphery. By sinus injection the venæ auricular. post. are most rapidly filled from the emissarium mast.; the venæ auricular. ant. can be filled from the vena temporal. superficial., which must be looked for on the temple, and injected towards the centre. By this means the plexus venosus pterygoid. in the fossa intratemporal. is frequently also injected, and thence the fluid passes into the venæ auricular. profundæ. Should the latter plexus have failed to be filled by the injection through the venæ jugular. communes et temporales superficiales, it is safest to fill it from the vena temporalis media. This vein is to be found above the zygomatic arch in the layer of fat between the superficial and deep lamina of the fascia temporal., and when discovered it is cut into, ligatured towards the periphery, and injected downwards. By injecting the vena facial. ant., which is to be looked for at the anterior margin of the insertion of the masseter, the plexus pterygoid. may, as a rule, also be filled through the vena anastomotica facialis, which leads out of this plexus into the vena facialis antica.

Heads from older individuals are better adapted for the injection of the veins than those of younger ones, because the valves at a maturer age become not infrequently insufficient, and the venous walls possessing greater firmness, extravasation does not so easily occur as in younger subjects.

The injection of the lymphatic vessels of the external region of the ear is effected either by means of metallic quicksilver or with soluble Berlin blue. In the latter case it is preferable to make use of a well-working Pravaz syringe with a finely pointed canula, which is inserted obliquely into the corium of the concha close to the orifice of the ear, after which the injection is proceeded with by gradually-increasing pressure. The quicksilver finds its way into the interstices of the corium, and from here into the lymphatic vessels. The injection will fail if the point of the canula be forced into the subcutaneous connective tissue, in which case numerous extravasations will occur in the meshes of the tissue. The anatomical museum of Paris contains several specimens of the auricle, in which the lymphatic vessels have been successfully injected with quicksilver.

PREPARATION OF THE NERVES OF THE ORGAN OF HEARING.

Preparing the nerves of the organ of hearing in the fresh state, when their finer branches are soft and liable to be torn, is much more difficult than when, by the action of some chemical agent, they have acquired the requisite degree of firmness and resistance. For this purpose Dr. Dalla Rosa, of the anatomical institute of Vienna, injects the head, separated from the trunk, for several days from the common carotids with a 0.5% aqueous solution of chromic acid, allowing 2-3 litres of this liquid to flow daily from an irrigator, placed about $1\frac{1}{2}$ m. above the head, which after each injection is immersed in the same solution, the liquid being changed every day. After a few days the head is placed in water, which is either kept running or changed several times in the day, and where it is allowed to soak for two or three days, during which time water is also allowed to flow repeatedly through the carotids from the irrigator. Finally the head is injected by means of an irrigator with 2-3 litres of 50% alcohol, and, after removal of the brain, preserved for the purpose of preparation in alcohol of the same concentration.

PREPARATION OF THE NERVUS AURICULARIS MAGNUS.

To represent this nerve the upper part of the platysma myoides is prepared. In the middle of the posterior edge of the musculus sterno-cleido-mastoid., the nerve comes into view, passes across the muscle obliquely forwards and upwards, and divides into two branches: the anterior ramus auricularis, and the posterior ramus mastoid. The latter accompanies the arteria auric. post. and radiates in the integument of the mastoid process and of the upper part of the auricle. The anterior branch passes on to the posterior surface of the lobe of the ear, and there gives off the cutaneous twigs for the posterior and (by means of perforating branches) the anterior surface of the lower half of the auricle, as well as for the external meatus. The inferior portion of the nervus auricul. magn. is covered by the platysma, and therefore does not become visible until this is removed.

PREPARATION OF THE TEMPORAL BRANCHES OF THE NERVUS AURICULO-TEMPORALIS.

The preparation of this nerve may be commenced either with that of the peripheral branches on the external region of the ear, or with the laying bare of the central portion of the nerve. In the former case a vertical incision is made in the skin in front of the tragus, and from here the integument of the temple is dissected off, leaving the subcutaneous connective tissue in position. Thus the branch of the auriculo-temporalis, ascending above the zygomatic arch, may be exposed, and its smaller branches, ramifying in the integument of the auricle and towards the external meatus, may be followed. The anastomoses of the nervus auriculo-temporalis with the nervus facialis are met with after dissecting away the upper part of the parotid gland, in the deeper layers between the maxillary joint and the anterior wall of the meatus.

PREPARATION OF THE ORIGIN OF THE NERVUS AURICULO-TEMPORALIS AND OF THE CHORDA TYMPANI.

In order to find the origin of the nervus auriculo-temporalis, the same proceeding is followed as in laying bare the arteria maxill. int. and the art. auric. prof., passing off from it. After removal of the musculus pterygoid. ext., together with the upper half of the ramus of the inferior maxilla, the third branch of the trigeminus, at its exit from the foramen ovale, is exposed. Immediately below this are found the two peduncular roots of the nervus auriculo-temporalis, which have between them the arteria mening. media before its entrance into the foram. spinos. and which, laterally from here, unite in a single trunk, passing outwards through the upper part of the parotid gland, and ascending, between the maxillary joint and the anterior wall of the meatus, above the zygomatic arch to the temple.

The chorda tymp. may be prepared along with the origin of the nervus auriculo-temporalis. Should it be desired to expose the nerve from the outside, it may be found at its exit from the Glaserian fissure immediately behind the fossa glenoidalis, or, what is easier, it may be traced from its junction with the trunk of the nervus lingualis upwards and backwards as far as the fissura Glaseri. In the same manner the preparation of the chorda tymp. can be effected, together with that of the otic ganglion, in following it, as shown in

Fig. 108, from its anastomosis with the lingual nerve as far as the tympanic cavity, and thence into the facial canal.

PREPARATION OF THE GANGLION OTICUM.

The representation of the otic ganglion is one of the most difficult preparations. It is made from within, on a skull cut vertically into halves, as the ganglion is situated on the medial side of the third branch of the trigeminus (Fig. 108, go.).

In order to make the principal trunk of this branch accessible, the body of the sphenoid bone, together with the basilar portion of the occipital bone, and the cartilaginous Eustachian tube with the levator palati mollis, are removed; the tensor palati, however, in its greater part, is retained. The principal trunk of the third division of the fifth nerve having thus been laid bare, the exposure of the ganglion may now begin. For this purpose, the nervus pterygoideus internus must be looked for, as it forms an important guide for finding it. With this object, after cutting away the remainder of the tongue, the inner side of the musculus pterygoideus is dissected clean, and the nerve sought for where it enters at the upper margin of that muscle, following it in a central direction towards the principal trunk. By this proceeding, the ganglion cannot escape the notice of the dissector, since the central termination of the nervus pterygoideus internus, entering the main trunk of the nerve, pierces the ganglion. The preparation will make it plain that the latter lies immediately below the foramen ovale, and is surrounded by vascular cellular tissue. In dissecting this away, beginners, taking it for adipose tissue, are apt to remove the ganglion along with it, which makes it advisable to practise at first on objects hardened in alcohol. The preparatory work is carried out as far as the excision of the body of the sphenoid, of the pars basilaris, and of the Eustachian tube; the nervus pterygoideus internus remains to be exposed. After this the object is first well steeped in water (one to two days), next hardened in alcohol, and not until then should the preparation of the ganglion be commenced.

The branches of the ganglion radiate from its anterior and posterior pole. Of the two posterior nerves, the inferior belongs to the tensor tympani; the superior is termed nervus petrosus superficialis minor. The anterior branches supply the tensor palati mollis.

The preparation of the nervus petrosus superficialis minor is effected from its origin at the otic ganglion through the fissura spheno-petrosa. Having passed through the latter, it runs laterally and parallel to the nervus petrosus superficialis major through the canaliculus tymp., at the superior opening of which it enters the tympanic cavity to anasto- mose with Jacobson's nerve.

PREPARATION OF THE RAMUS AURICULARIS VAGI.

The preparation of the ramus auricularis vagi is best commenced by exposing its peripheral portion. For this purpose, the posterior surface of the auricle is carefully dissected away from its insertion at the mastoid process, in order to find the peripheral portion, which runs in the stratum of connective tissue between the auricle and the

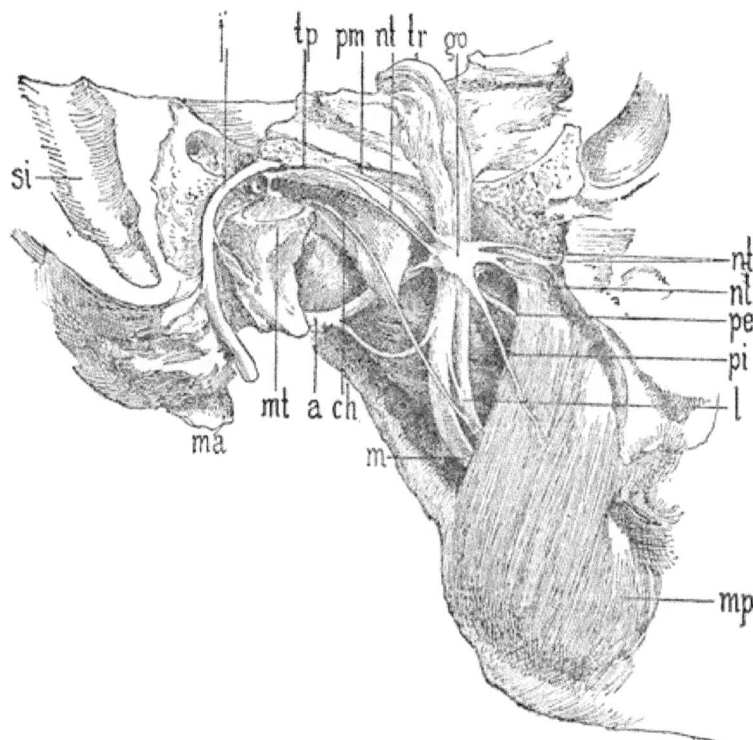

Fig. 108.—Preparation of the otic ganglion and the nerve branches issuing from it.—tr = nervus trigeminus with the ganglion Gasseri ; go = ganglion oticum ; nt′ = nervus tens. palat. moll. ; nl = nervus levat. palat. moll. ; pe = nervus pterygoid. ext. ; pi = nervus pterygoid. int. ; tp = musc. tens. tymp. ; mt = nervus tens. tympani ; pm = nervus petrosus superf. minor ; m = nervus mandibularis ; ch = chorda tymp. : l = nervus lingualis and its anastomosis with the chorda tymp., forming an acute angle ; a = nervus auriculo-tempor. ; f = nervus facialis ; mt = inner surface of the membrana tympani ; ma = proc. mast. ; si = sinus transvers. ; mp = musculus pterygoid. int.   After a preparation in the anatomical museum of Vienna.

opening of the canaliculus mastoid., and enters the fissura tympanico-mastoidea.

To prepare the nerve from its origin, the pars condyloidea of the occiput is removed, the nervus vagus then followed into the fossa jugularis, where the ganglion jugulare is laid bare, when the ramus

auricularis vagi, passing off from behind the latter, can be searched
for.

The most difficult part of the preparation of this nerve is to follow.
It consists in opening with the chisel the canaliculus mastoid. from
the posterior part of the fossa jugularis to the canalis Fallopii, in
which the nerve anastomoses with the facial nerve, and from here to
its opening at the fissura tympanico-mastoidea. The preparation of
this nerve in its whole extent is rarely successful, even after repeated
attempts, and it is therefore advisable first to open the canaliculus
mastoid. on the macerated temporal bone.

PREPARATION OF THE NERVUS ACUSTICUS AND NERVUS FACIALIS.

To represent the nervus acusticus in its whole length, after removal
of the cerebrum and dividing the tentorium cerebelli at the edge of

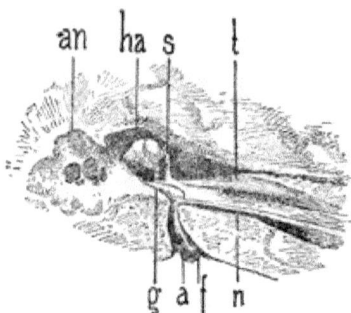

Fig. 109.—View of the tympanic cavity after the removal of the tegmen tymp. (Right
ear.)—ha = malleo-incudal articulation ; t = musc. tens. tymp. ; s = tendon of the musc.
tens. tymp., passing across the tympanum ; f = nerv. facialis ; g = genu nervi facialis ;
n = nerv. petros. superf. major ; a = nerv. acusticus ; an = antrum mast.   After a
preparation in my collection.

the pyramid, the cerebellum, together with the medulla, is raised up a
little, and the acoustic nerve cut through, close to its point of exit
at the side of the medulla oblongata, along with the neighbouring
facial nerve.   Next, the upper wall of the internal meatus is removed
with chisel and forceps as far as its floor, then the fibrous lining of
the canal, a continuation of the dura mater, is split, and the acoustic
nerve, running in company with the facial, will be exposed.   The
latter, recognisable by its firmer consistence (portio dura), lies above
the acusticus.   This nerve, somewhat softer (portio mollis), divides, at
the bottom of the meatus, into an anterior branch which enters the
modiolus and the lower convolution of the cochlea (ramus cochleæ),
and a posterior branch, penetrating the bone between the cochlea and
vestibule, and passing into the latter (ramus vestibuli) ; before its

entrance it shows a slight swelling (intumescentia ganglioformis Scarpæ), and a small branch passes from it to the inferior ampulla.

The preparation of the nervus facialis is carried out along with that of the acoustic nerve. The upper wall of the internal meatus having been removed, and the trunks of the two nerves exposed, as far as the fundus, the object is fixed in the vice, and the osseous mass situated above the vestibule and the cochlea cut away with the chisel (4-5 mm. in front of the superior semicircular canal), and that portion of the facial nerve, which passes transversely from within outwards through the petrous bone, is laid bare up to the genu nervi facialis at the boundary of the inner wall of the tympanic cavity. In order to prepare the nerve in its further course in the tympanic cavity, the pars tympanica et squamosa must be separated from the pyramid and the mastoid portion in such a manner that the posterior lower part of the pars tymp., lying within the foramen stylomast., remains connected with the pars petrosa. The preparation of the nerve, from the genu to the descending portion of the facial canal, is the easiest part of the work, as this part of the canal, with its thin, at times dehiscent osseous wall, projects above the niche of the fenestra ovalis as an elevation, descending obliquely in a direction backwards and downwards, and a slight pressure with a broad graver will be sufficient to remove the thin osseous lamella.

Somewhat more elaborate is the exposure of the descending portion of the facial nerve, because it is covered by a thicker layer of bone, the removal of which with the chisel requires more time. The direction in which the preparation must be proceeded with is given by the point at the posterior tympanic wall, where the horizontal meets the descending portion at an obtuse angle, and by the exit of the nerve from the foramen stylo-mastoideum.

The representation of the course of the facial nerve is materially facilitated on decalcified petrous bones. Here it suffices to remove in layers, with a scalpel, the superior wall of the internal meatus. simultaneously with the bone on the upper surface of the pars petrosa, covering the facialis, to speedily come upon the nerve. In a like manner its horizontal and descending piece in the tympanic cavity is also quickly brought to view. Such preparations, preserved in spirits of wine, are specially adapted for purposes of demonstration.

Among the nerves communicating with the nervus facialis, the nervus petros. superf. major may be mentioned. Its preparation is very easy, since it occupies a groove in the dura mater, and runs, as described on p. 28, from the anterior angle of the temporal bone to the hiat. canal. Fallop. on the upper surface of the pars petrosa; its course as far as the genu can, therefore, be represented with pincette

10

and scalpel without further manipulation on the bone. The direction of its course corresponds, as the accompanying illustration (Fig. 109) shows, with that of the musculus tensor tymp.

The question as to how the anastomosis of the nervus petr. sup. major with the facialis takes place, has been decided by Frühwald (Sitzungsber. d. kais. Acad. d. Wiss. in Wien, 1876, Bd. 74), who has shown that the facial nerve gives off fibres to the nervus petr. sup. major, but that, on the other hand, it receives fibres from the nervus petr. major. As the method of examination suggested by Frühwald may also be employed to demonstrate the course of other nerve bundles, it shall be briefly described here: The nerve to be placed in 95% alcohol, until it loses its natural softness, and then saturated in oil of turpentine. In order to destroy the connective tissue, the preparation is heated on a water-bath in 50 ccm. of a concentrated solution of chlorate of potash and $3\frac{1}{2}$ ccm. of fuming nitric acid, until the nerve becomes soft again. After boiling, it is to be placed for one or two days in distilled water, and after that in equal parts of a concentrated solution of tartaric acid and glycerine, which clears up the remaining connective tissue, leaving, however, the nerve fibres white. The latter are examined on a black ground of wax.

The exposure of that portion of the chorda tympani, which runs in the canaliculus chordæ on the posterior tympanic wall as far as the facial canal, is difficult, on account of the narrowness of the small furrow, and its variable position in the osseous mass behind the membrana tympani, especially if it be desired to show both nerve and chorda in their continuity. In opening the narrow canal with the chisel, injury to the thread-like chorda can scarcely be avoided, and it is therefore laid bare more rapidly on decalcified preparations.

In addition, a method of opening the entire facial canal on the macerated temporal bone, as suggested by Kiesselbach, may be noted. He commences the cut in a line which unites the middle of the hiatus aquæd. vestib. on the posterior surface of the pyramid and the posterior edge of the foram. stylo-mastoid., and meets the proc. mastoideus. The cut then passes towards the posterior margin of the internal meatus, and after that in a plane, which lies through the opening at the commencement of the facial canal and the spina supra meat.; it next turns towards the upper edge of the genu facial., and thence in any direction through the cochlea. The surface of the cut anteriorly shows, besides the whole of the canalis Fallop., the lower part of the vestibule with the foram. ovale and the commencement of the cochlea; posteriorly, the position of the antr. mast. and the upper portion of the vestibule with the semicircular canals.

14. MAKING TOPOGRAPHICAL SECTIONS OF THE ORGAN OF HEARING
FOR INSTRUCTION PURPOSES.

For the study of the topography of the organ of hearing, in addition to the previously described preparations, a series of sections of the entire temporal bone is required, the most important of which shall here be discussed.

The topographical preparations of the ear in my collection are divided into: 1, sections of common preparations of the temporal bone, preserved in spirits of wine; 2, sections of decalcified temporal bones; 3, topographical sections on dry preparations.

(a) Topographical Sections of the Temporal Bone prepared for Preservation in Spirits of Wine.

The technical proceeding in making topographical preparations of the ear, which are to be preserved in spirits of wine, is the following On a specimen which has been soaked in water and sufficiently hardened in alcohol, the external soft parts, auricle, and cartilaginous meatus, are first divided by means of a sharp scalpel, in the direction in which the cut is to be made, if possible at one sweep down to the periosteum, the cut being continued through the bone in the same direction with a fret-saw. On preparations where it passes through the membrani tympani, the latter must be divided with a small pointed knife from the external meatus (under illumination with the mirror), and exactly in the direction of the saw-cut.

A more exact representation of the topographical position of the external soft parts, in their relationship to the cavities in the temporal bone, is obtained by sections on frozen specimens. These admit of dividing the soft parts and bone with the saw in any direction without great difficulty, while the displacement of the tissues, frequently occurring in ordinary spirit preparations, is avoided. The freezing of the preparation may be effected either by exposing it to the open air during frosty weather, or by artificial means. In the latter case the object, simply wrapped in a piece of linen, is surrounded by a mixture of two-thirds snow or pounded ice, and one-third common salt, and allowed to stand for two or three hours in the cold open air, or in a cool place. The frozen sections are quickly washed, and immediately placed in concentrated spirit, which must be changed after twenty-four hours, whereby the soft parts retain their firmness and original positions.

The topographical preparations of this kind most important for the purposes of instruction are the following :

1. A horizontal section through the entire temporal bone, so

carried through the auricle that the plane of the cut coincides with
that part of the antrum mastoid. which is situated behind the cavum.
tymp. (Fig. 110).

In order to fix accurately the direction of the cut, we first remove
the tegmen tymp. and mastoid. from a temporal bone, sawn out
in the regular way, and with the auricle *in situ*, so as to obtain a
view into the antrum mastoid.   Then, placing the point of one leg
of a pair of compasses in the antrum, and the other on a corresponding place on the concha, the spot is determined through which the
horizontal cut is to be carried.   Now, this spot on the anterior surface
of the auricle is either in the portion of the concha bordering on the
mastoid process, behind the upper margin of the external orifice, or
somewhat higher up, above the transverse ridge of the concha, formed

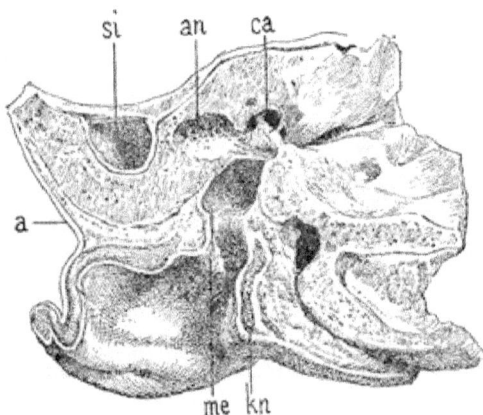

Fig. 110.—Horizontal section through the temporal bone at the level of the antrum mastoid.
—a = insertion of the posterior surface of the auricle at the mastoid process ; me =
meatus audit. ext. ; kn = section of the tragus and of the anterior cartilaginous wall of
the meatus ; ca = cavum tympani with the ossicles ; an = antrum mastoid. ; si = sinus
transversus.   After a preparation in my collection.

by the continuation of the antihelix.   In making such sections,
numerous exceptions are met with, due partly to the strong arching
forwards and outwards of the sinus transversus, partly to an
abnormally high or low position of the auricle.

To make these horizontal sections, the auricle is divided, at a point
determined by the compasses, with a sharp scalpel, down to the bone
(on frozen preparations it is sawn through).   It will now be better to
saw the bone through in the direction of the incision in the soft parts
from behind forwards, whereby a horizontal section is obtained, on
which, as the accompanying illustration shows, the relative positions
of the external and middle ear come clearly into view.   To avoid
injuring the malleus and incus while proceeding with the section, the

plane of the cut must, at the point where the antrum passes into the tympanic cavity, lie a little higher.

The horizontal section through the temporal bone here described is suitable for studying the external meatus in its course from the orifice to the tympanic membrane. A complete representation of the angular curvatures of the canal is, however, only to be obtained by placing side by side horizontal and frontal sections (Figs. 110 and 111). These further show the mode of apposition of the posterior cartilaginous wall of the meatus to the anterior surface of the mastoid process, the relationship of the anterior wall of the meatus to the maxillary joint, as well as that of the attachment of the posterior

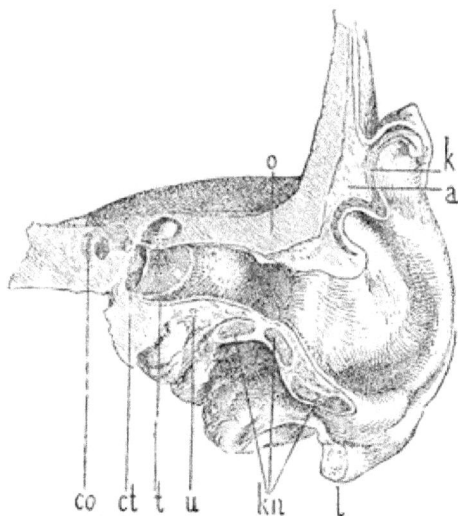

FIG. 111.—Frontal section through the temporal bone.—k = upper portion of the cartilage of the auricle ; a = insertion of the auricle in the temporal region ; l = lobulus ; kn = sections of the lower portion of the auricle and of the inferior cartilaginous wall of the meatus ; o = superior, u = inferior, wall of the osseous meatus ; t = membrana tymp. ; ct = cavum tymp. ; co = cochlea.

surface of the auricle to the mastoid process, which is of importance in the surgical opening of the latter. It is only by the examination of such preparations that we recognise the necessity of a partial detachment of the auricle from its insertion, in order, during operative interference, to reach the antrum mastoid. by the shortest way.

2. A frontal section through the auricle, external meatus, membrana tympani, tympanic cavity, and labyrinth.

The cut takes such a direction as to pass through the middle of the outer opening of the ear and through the anterior portion of the lobe. After dividing the auricle, fasciæ, and temporal muscle down to the bone, the incision is continued with a small scalpel through the upper

and lower cartilagino-membranous walls of the meatus to the union
with its osseous portion.   After this the cut is carried farther with a
fret-saw through the upper and lower walls of the osseous meatus to
the vicinity of the membrana tympani.   To protect the latter from
being torn by the saw, it is, under sufficient illumination of the external meatus, cut through with a small knife close in front of the
manubrium from above downwards, and the saw-cut continued in such
a manner that it passes in front of the malleus towards the tympanic
cavity and labyrinth.

Such frontal sections (Fig. 111) of the organ of hearing demonstrate the mode of apposition of the upper parts of the auricle (k)
to the temporal muscle, the relation of the superior membranous wall

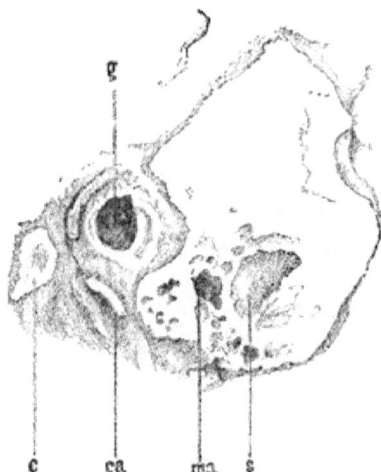

FIG. 112.—Sagittal section through the temporal bone, 2 mm. inwardly from the external
     orifice of the ear.—g = section of the cartilagino-membranous meatus ; ca = section of
     the lower wall of the cartilaginous groove ; ma = mastoid cells ; s = sinus transvers.,
     opened ; c = section of the outer extremity of the proc. condyloid. of the inferior maxilla.

of the meatus to the horizontal part of the squama, the direction of
the course of the external meatus in the frontal plane, and the relationship of its lower wall to the parotis.

3. A series of sagittal sections through the external meatus, mastoid
process, and maxillary joint.

It is best to use for this purpose firmly-frozen preparations, on
which the auricle and cartilaginous meatus, as well as the ascending
ramus of the inferior maxilla, have been left intact *in situ*.   The first
cut (Fig. 112), 2 mm. inwardly from the external orifice, is carried
through the cartilagino-membranous meatus (g), the outer part of the
mastoid process (ma), and meets the outermost end of the proc. condyloideus of the ramus of the inferior maxilla (c).   This section shows,

in addition to the relative position of the maxillary joint to the meatus, the mode of apposition of the cartilagino-membranous meatus to the

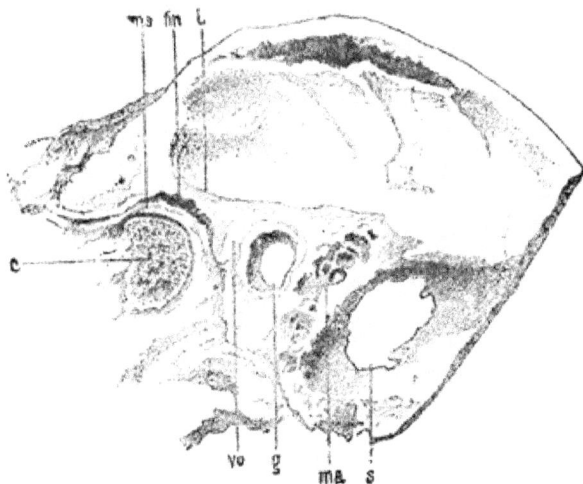

FIG. 113.—Sagittal section through the temporal bone, 4 mm. medially from the previous one.—g = lumen of the osseous meatus; vo = its anterior wall; ma = mastoid cells; s = sinus transvers., opened; c = section of the proc. condyloid. of the inferior maxilla; me = meniscus of the maxillary articulation; fm = fossa mandibularis.

anterior surface of the mastoid process. If the sinus transversus (s) be strongly arched outwards, it will be exposed by this cut.

A second sagittal cut (Fig. 113), made 4 mm. medially, divides the

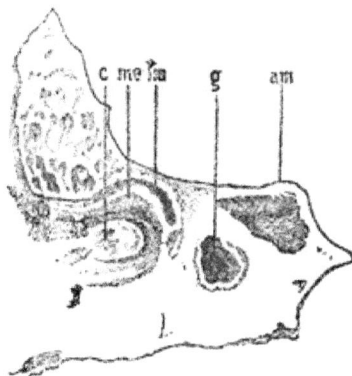

FIG. 114.—Sagittal section through the temporal bone at the inner end of the osseous meatus.—g = osseous meatus; am = antrum mast.; c = section of the medial extremity of the proc. condyloid. of the ramus of the inferior maxilla; me = meniscus; fm = fossa mandibularis.

osseous meatus (g), the articular fossa of the lower jaw (fm), the entire condyle (c), and its meniscus (me). The space between the maxillary joint, the auditory meatus, and mastoid process is occupied by parotis and adipose tissue. Besides the exceedingly variable number of the mastoid cells, there are frequently found on this section the outermost space of the antr. mastoid. (ma), and a considerable gap in the sinus transversus.

Lastly, a third sagittal section through the osseous meatus, immediately in front of the upper pole of the membrana tympani (Fig. 114), shows, besides the elliptical opening of the section of the osseous meatus (g), the transverse section of the inner extremity of the proc. condyloideus of the inferior maxilla, with the articular fossa (fm), the meniscus (me), and, behind, the transverse section of the antr. mastoid. (am) in its greatest extent.

4. A frontal section through the skull, to show the relative position of the two cartilaginous Eustachian tubes at the base of the skull.

Most suitable for this is a firmly frozen head, which, while in the normal position, is sawn through in a frontal plane, passing through both oval windows. The posterior view of this section (Fig. 115) shows the middle portion of the naso-pharyngeal space (R), on the lateral wall of which the medial plate of the cartilage of the tube (t), is symmetrically placed. The latter, standing obliquely to the base of the skull, and connected with the basilar fibro-cartillage (f), is bent in an upward direction, and forms the lateral cartilage hook (h h'), which roofs over the slit of the tube (sp sp'), directed almost perpendicularly downwards. In addition to this view of the Eustachian tubes, the section shows the topographical position of the tensor (ts) and levator (l) palati mollis, of the m. pterygoid. int. (pi) et ext. (pe). As a matter of course the aspect of this section of the tube and the size of the slit will vary according to whether the plane of the cut approaches the ostium pharyng. or the ost. tymp.

In order, therefore, to obtain a clear representation of the section of the tube, it is advisable to carry at least three parallel frontal cuts through the base of the skull, so that the first falls immediately behind the ostium pharyng. tubæ, while the second passes through the cartilaginous tubes in the plane of the foramen ovale, and the third, 4-5 mm. further back, in a frontal direction through the osseous tubes.

5. A frontal section through the entire skull and brain in the frontal plane of the two external auditory canals.

The favourable results recently obtained by Thomas Barr and William McEwen, of Glasgow, in their operative treatment of otitic cerebral abscesses, must be regarded as a fresh and most important advance in dealing with dangerous affections of the ear.

For such operations, an accurate knowledge of the topographical relations of those parts of the brain, which are situated above the external meatus and the middle ear, is therefore of the greatest importance.

With the object of ascertaining the extent and size of the temporal lobe, in which cerebral abscesses most frequently have their seat, the whole skull and brain must be so divided in a frontal direction that the plane of the section passes on both sides through the external meatus and the tegmen tymp. For this purpose it is preferable to use the entire head of a middle-aged individual, which, if there be no

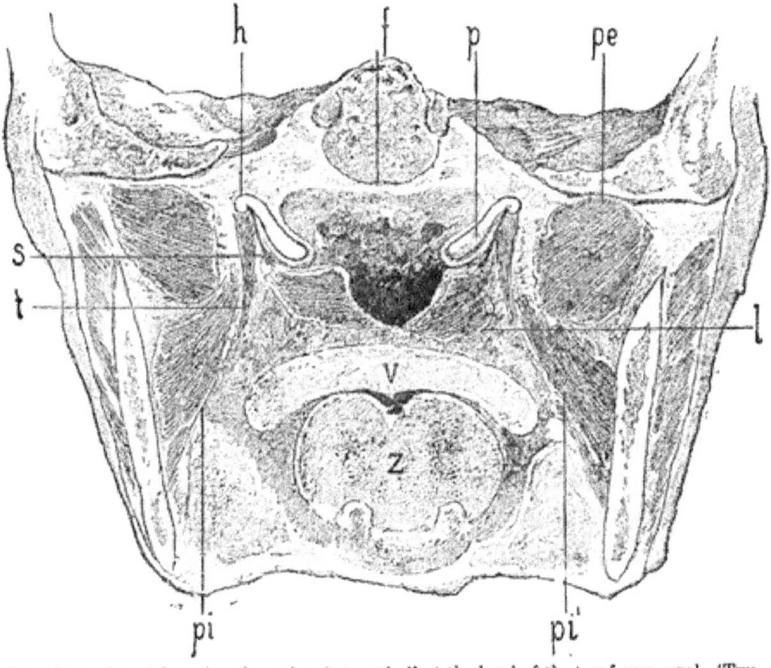

FIG. 115.—Frontal section through a frozen skull at the level of the two foram. oval. (Two-thirds of its natural size.)—f = fibro-cartilago basilaris ; p = section of the medial cartilage plate ; h = lateral cartilage hook of the Eustachian tube ; s = slit-like opening of the tube ; l = section of the musc. levat. palat. mollis ; t = section of the musc. tensor. palat. mollis ; pe = musc. pterygoid. extern. ; pi pi' = musc. pterygoid. intern. ; v = palat. molle ; z = tongue.

frost, is made to freeze in a mixture consisting of two-thirds snow or pounded ice, and one-third common salt. In from eight to ten hours the whole head will be so firmly frozen that it can be sawn through with an ordinary joiner's saw, while the brain retains its original position in the skull. In order not to miss the direction of the cut, it is well to mark it on the vertex of the head, after the hair has been removed, by a frontal incision passing from the outer opening of one ear to that of the other.

When this cut has been made, a second one is carried through the skull 2-2½ cm. behind, and parallel to the first, for the purpose of reducing the bulk of the object for preservation. The bone-dust adhering to the surface is scraped off with a long knife, the preparation quickly washed in iced water, and immediately placed in concentrated alcohol. After twenty-four hours it is transferred to fresh concentrated spirit, in which the object, now quite freed from water, retains its firm consistency. Commonly, this method of preparation results in the shrinking of the brain *in toto*, caused by the action of the alcohol. As, thereby, its various parts are drawn somewhat towards the middle, this should be taken into account when estimating the topographical position of the different portions of the temporal lobe as regards their relationship to the organ of hearing.

Topographical sections of the brain may also be made on heads hardened in alcohol, the use of the saw, however, being inadmissible. The proceeding is the following: As in the anatomical opening of the cranial cavity, the roof of the skull is sawn through from one auricle to the other, in the plane before mentioned, down to the dura mater; this is then divided in the direction of the saw-cut with a long, thin, and sharp knife from above downwards to the base of the skull, and finally the cut completed through the latter with a fine saw.

A clear representation of the more exact topographical relations of the temporal lobe to the temporal bone is obtained by making several parallel frontal cuts through the entire skull, the most anterior of which passes through the apex of the pyramid, the middle one through the tympanic cavity, and the posterior one through the antrum mastoid. The accompanying illustration (Fig. 116) is taken from a preparation in which, on the left side, the external meatus and tympanic cavity, on the right side the antrum mastoid, and the posterior portion of the vestibule with the semicircular canals, were met. We notice in this section, above the osseous meatus, the gyrus inferior (gi) of the temporal lobe, which is conspicuous at the transition of the squama into the horizontal portion of the middle cranial fossa, by a strongly-marked impressio digitata. Farther inwards, corresponding to the tympanic cavity (t), are situated the most anterior portions of the gyrus fusiformis (the gyrus lying between gi and fh), which in part extends beyond the labyrinth and receives a deep impression (impressio petrosa) through the eminentia arcuata of the upper surface of the pyramid. Finally, with the medial and most anterior portion of the pyramid corresponds the gyrus hippocampi (gh), which laterally adjoins the tentorium cerebelli.

To reach otitic cerebral abscesses by operation, it is of importance to know the diameters of the temporal lobe above the middle ear, and measurements noted by me on several preparations are given below. According to these, the transverse diameter of the temporal lobe immediately above the external meatus, that is, from the lateral edge of the cerebral hemisphere to the medial border of the gyrus hippo-

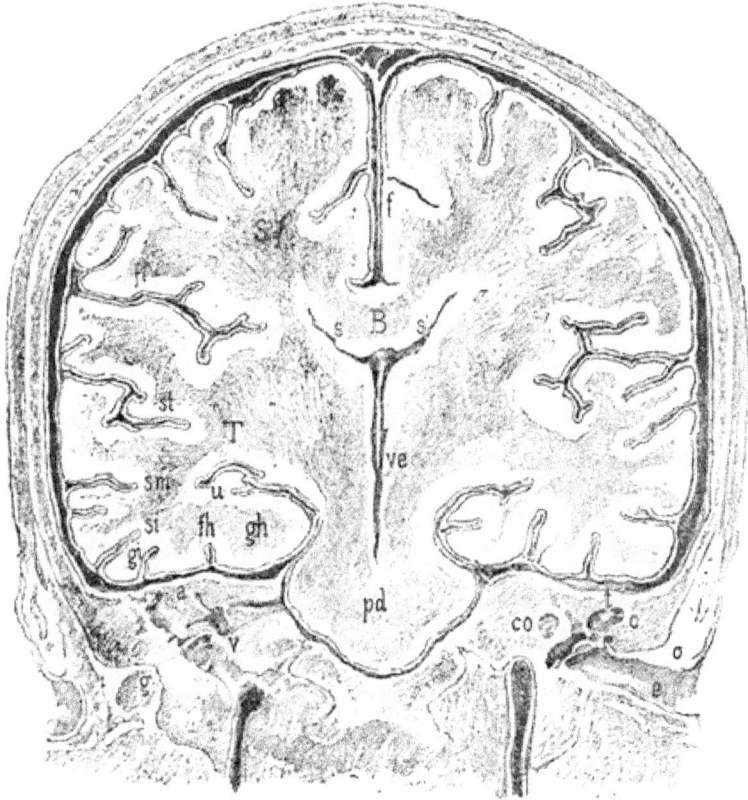

Fig. 116.—Frontal section through the frozen skull and brain : on the left through the vertical plane of the external meatus and of the tympanum ; on the right through the mastoid antrum.—On the left: e = meat. aud. extern. ; o = upper wall of the osseous meatus ; c = cavum tympani ; t = tegmen tymp. ; co = cochlea. On the right : a = antrum mastoid. ; v = vestibule with portions of the superior and horizontal semicircular canals ; T = temporal lobe ; S = parietal lobe ; f = falx duræ matris ; s s' = lateral ventricle ; v = third ventricle ; B = pons ; pd = pedunculus cerebri ; fi = fissura Sylvii ; st = sulcus temp. super. ; sm = sulcus temp. med. ; si = sulcus temp. infer. ; gi = gyrus temp. inf. ; gi-fh = gyrus fusiformis ; fh = fissura hippocampi ; gh = gyrus hippocampi ; u = inferior cornu.

campi, amounts to 5 cm. The breadth of a section of the brain made 4 cm. higher, namely, on a level with the fissur. tempor. sup. from the external surface of the temporal lobe to the third ventricle, measures

6-6½ cm. The transverse diameter of the hemisphere on a level with the fissura Sylvii to the lateral ventricle amounts to 5 cm. The height of the temporal lobe above the tympanic cavity measures 4-4½ cm., that of the temporo-parietal lobe 7-7½ cm.

(b) Topographical Sections of the Organ of Hearing on Decalcified Temporal Bones.

This method of preparation offers such great advantages for the study of the topography of the organ of hearing, as to make it a matter of surprise that, up to the present, it has not been in use to a greater extent in the anatomical institutions. For the decalcification of the entire temporal bone, either 5-10% hydrochloric acid, or a mixture consisting of 100 parts water, 3½% nitric acid, and ¾% common salt may be used. The quantity of the fluid necessary for uniform decalcification of one or two temporal bones varies from 1-1½ litres, and must be renewed at least every second day. The process, which occupies two to three weeks, may be considered complete if a needle, thrust into the pyramid behind and below the porus acust. int., does not meet with any resistance (Moos). The decalcified preparation, when freed from any acid remaining in it, either by placing it for several hours under a stream of water, or in a weak alkaline solution, is finally soaked for several weeks in 50-60% alcohol for the purpose of hardening it. If placed in rectified spirit, the decalcified bone becomes, after a short time, so hard that it can only be cut through imperfectly and with great difficulty.

Sections may now be made, without any further preparation, on decalcified objects taken direct out of the spirit, or on such as have been decalcified and saturated with celloidin. The latter method (see 'Celloidin Embedding' in the Histological part) is especially suitable when, on the topographical section, the vestibular saccules, the ampullæ, semicircular canals, and membranous cochlea are to be retained in their positions. Where the celloidin, on a section, interferes with the clearness of the details, it may be dissolved by placing it in a mixture of alcohol and ether. With careful handling, the structures of the labyrinth do not suffer thereby any change of position, particularly if, before decalcification, the preparation had been placed in a suitable fixing and hardening fluid. (Compare the section 'Preparatory Methods' in the Histological part.)

In making sections on decalcified temporal bones, a long, narrow, flat knife is used, with which, if the cut is to be made in one plane only, the incision, where possible, is carried through the preparation at one sweep, the direction of the cut having been previously marked. On preparations, however, where various parts of the organ not lying in one plane are to be brought into view, the cut must be made at intervals.

It would lead too far to enumerate all the sections that might possibly be made on decalcified temporal bones. It need here only be noted that, both on frontal and horizontal sections through the entire temporal bone, the topographical position of the various parts of the auditory organ can, in many respects, be brought much more clearly into view than on frozen preparations. Thus, while in the sawing through of temporal bones which have not been decalcified, the membrana tympani may frequently be injured, the ossicula dislocated, and the labyrinthine structures destroyed, the cut can, on decalcified preparations, be made in any plane passing through the tympanic membrane, ossicles and labyrinth, without running the risk of injuring important structures, and thus spoiling the anatomical picture. As especially instructive preparations of this kind we here mention :

1. Three horizontal sections through the temporal bone, the first of which passes through the upper wall of the osseous meatus, the upper tympanic space, the malleo-incudal articulation, the upper periphery of the cochlea, and the superior semicircular canal. The second, somewhat lower, passes through the external meatus, immediately below its superior osseous wall, divides the malleus above its short process and above the insertion of the tendon of the tensor tymp., meets the cochlea in its greatest diameter, and opens the upper part of the vestibule. The third cut, lower still, lies in the axis of the osseous meatus, divides the inferior extremity of the manubrium, the incudo-stapedial connection, and the foot-plate of the stapes in the fenestra ovalis, meets the lower segment of the cochlea, and lays bare the vestibule in its greatest circumference.

2. Two or three frontal sections, the most anterior of which passes through the front of the tympanic cavity, immediately behind the ostium tympanicum tubæ, and divides the cochlea nearly in its axis. The second, lying 3-4 mm. behind, is best made at that spot where the umbo of the tympanic membrane approaches most closely the wall of the promontory. Particularly instructive is the frontal cut made in this region, if it passes exactly through the longitudinal axis of the malleus, which is thus divided from its head to the umbo into halves. Such sections give a general idea of the height and breadth of the tympanic cavity, the relative curvature of the membrana tympani and of the promontorial wall, as well as of the position of the ossicula in the tympanic space. A third frontal cut, 2-2½ mm. behind the second, meets the posterior segment of the tympanic membrane, the body of the incus and its long process, crosses the niches of the oval

and round windows, and exposes the posterior wall of the vestibule in its entire extent.

3. A frontal section through the membrana tympani in its connection with the malleus and incus, but detached from the pyramid for the demonstration of the relative positions of Shrapnell's membrane, of the system of cavities, situated between the malleus and incus and the lateral niche of the external tympanic wall, as well as of the ligamentous apparatus of the malleus and incus (see Fig. 69, p. 86). (For the more detailed method of preparation of this region see pp. 86 and 87.)

Two horizontal sections, the first of which divides the malleus and incus in a plane lying through the short process of the incus and the ligamentum mallei anterius, the second, somewhat lower, meeting the malleus and incus above the short process of the former, are indispensable as supplementary preparations for the study of the anatomical relations of this, in a pathological respect, so important region.

4. A series of horizontal and vertical sections through the pyramid, first decalcified, then dehydrated in alcohol, and embedded in celloidin, in order to show the topographical positions of the saccules, ampullæ, and semicircular canals, as well as of the membranous cochlea. As regards the direction to be observed while making these cuts, we refer to the former chapter, 'Preparation of the Membranous Labyrinth,' p. 127.

(c) Preparation of Topographical Sections of the Organ of Hearing by the Dry Process.

To make dry preparations of the organ of hearing, specimens preserved in spirits of wine are made use of, which are dried at an ordinary temperature in a drying apparatus, provided with several air-holes and protected from dust. This applies only to those preparations from which, with the exception of the membrana tympani, the intra-tympanic muscles, and the labyrinthine fenestræ, all the soft parts have been removed. Since, by the drying of the object, the membrana tympani becomes very easily fissured, it should, while still moist, be brushed over with a mixture of sublimate glycerine (3 parts glycerine, 7 parts water, 0.01 corrosive sublimate). By this means the membrane retains its pliancy, and by the addition of the sublimate it is preserved from being destroyed by the larvæ of insects.

With objects, however, where the soft parts—cartilage of the ear, cartilaginous meatus, Eustachian tube with its muscles, and the dura mater—are to be retained on the preparation, it should, after Professor Laskowski's excellent method, be saturated in a 5% solution of carbolized glycerine, then dried and sawn.

Of the topographical preparations of the organ of hearing the following sections are suitable for the purposes of study and demonstration :

1. A preparation on which the external surface of the membrana tympani, the anterior aspect of the tympanic cavity with the ossicular chain and the tendon of the tensor tymp., the vestibule, and internal auditory meatus are represented (Fig. 117). The mode of preparing is as follows :

After removal of the anterior, and in part also of the inferior osseous wall of the meatus, according to the method already described (p. 58), a frontal cut is made by means of the fret-saw through the middle of the superior wall of the osseous meatus, up to the vicinity of the tympanic membrane. Here the cut turns forwards and slightly

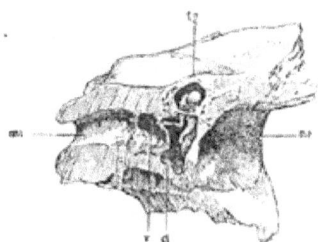

FIG. 117. — Frontal section through the external meatus, tympanum and the labyrinth of an adult. The annulus tymp. and the tympanic membrane are preserved intact. (Left ear.) — me = meat. audit. extern. ; ct = cavum tymp. with the tympanic membrane, the ossicular chain and the tensor tendon. ; tg = tegmen tymp. ; v = vestibule ; mi = meat. audit. intern. After a preparation in my collection.

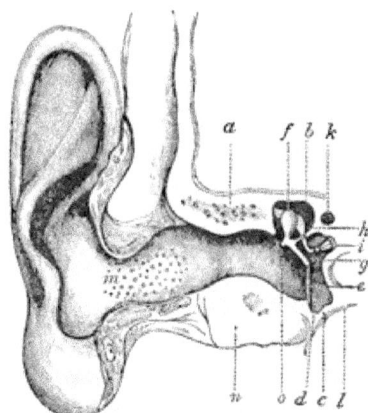

FIG. 118.—Frontal section through the external meatus, tympanic membrane and tympanum. (Right ear.)—a = cell spaces in the upper tympanic wall, communicating with the middle ear ; b = roof, c = lower wall of the tympanum ; e = tympanic membrane ; f = head of the malleus ; g = manubrium ; h = incus ; i = stapes ; k = canalis Fallopii ; l = fossa jugularis ; m = openings of glands in the external meatus.

downwards, and encircles the anterior periphery of the membrane 1-1½ mm. in front of the sulcus tymp., as far as the inferior wall of the tympanic cavity. By removing this piece of the external wall of the osseous Eustachian tube only an imperfect view into the tympanum is obtained. To lay bare its middle portion, together with the ossicular chain, the tegmen tymp. (tg) is cut through in a frontal direction immediately in front of the tendon of the tensor tymp., stretched transversely, and the saw-cut continued from here in front of the fenestra ovalis, through the vestibule (v) and the internal meatus (mi).

2. Equally instructive, especially for studying the mutual relations of the curvature of the membrana tympani and of the promontory, are preparations on which (Fig. 118), after frontally dividing the superior and inferior osseous walls of the meatus, the tympanic mem-

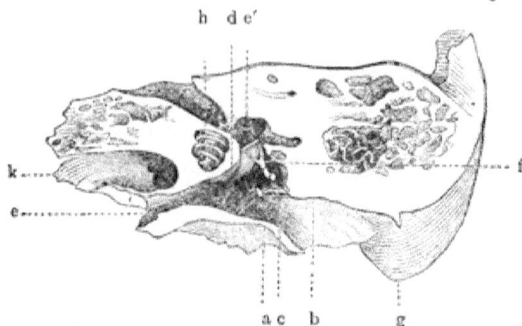

FIG. 119.—Horizontal section through the organ of hearing.—a = anterior wall of the osseous meatus; b = its posterior wall; c = section of the tympanic membrane, of the manubrium, and of the posterior pouch; d = promontory; e = ostium tymp. tubæ; f = stapes, in connection with the lower extremity of the long crus of the incus and the tendon of the stapedius; g = mastoid process; h = cochlea; i = vestibule; k = canalis caroticus.

brane is cut with a narrow, sharp knife immediately in front of the handle of the malleus down to its lower periphery, the course of the cut taking the same direction through the tympanic cavity and the labyrinth as in the preceding case.

3. A horizontal cut through the external meatus, tympanic cavity

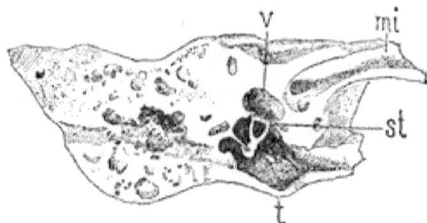

FIG. 120.—View into the tympanic cavity from below.—t = membrana tymp.; st = stapes, with the tendon of the musc. stapedius; v = vestibulum; mi = meat. audit. intern. After a preparation in my collection.

and labyrinth (Fig. 119). It runs in the external meatus just above its axis, meets the tympanic membrane (c) 2 mm. above the umbo, separates the manubrium and the lowest extremity of the long process of the incus, and proceeds through the labyrinth so as to open the vestibule (v) above the fenestra ovalis, and to expose the cochlea (h) with the modiolus and the membrana spiralis.

4. A horizontal section, on which the tympanic cavity and the labyrinth are laid bare from below (Fig. 120). After exposing the external surface of the tympanic membrane by removal of the anterior and lower walls of the meatus, the anterior wall and the floor of the tympanum are carefully broken away with the forceps; a cut is then made with the finest fret-saw round the lower and anterior periphery of the osseous frame of the tympanic membrane, the outer and inner surfaces of which become thus open to unobstructed examination. After this the pyramid is divided by a horizontal cut, which commences at the lower wall of the internal meatus, and terminates immediately below the fenestra ovalis at the inner wall of the tympanic cavity. By taking away the lower half of the pyramid, a general view

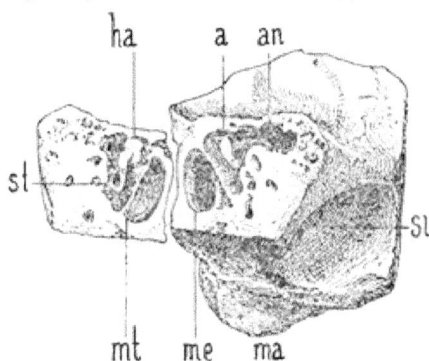

FIG. 121.—Frontal section through the innermost portion of the external meatus and the posterior part of the tympanic membrane.—me = meat. audit. ext.; mt = membran. tymp.; ha = malleo-incudal articulation, sawn through; a = surface of the saw-cut through the body of the incus; st = stapes; an = antrum mastoid.; si = sinus transv.; ma = proc. mastoid. After a preparation in my collection.

is obtained of the relative positions of the ossicles in the tympanic cavity, especially of the curvature of the membrana tympani, of the inclination of the lower extremity of the manubrium towards the inner tympanic wall, and of the incudo-stapedial connection.

5. A frontal cut, which meets the innermost portion of the osseous meatus, the posterior part of the membrana tympani and of the tympanic cavity (Fig. 121). Such sections show, on their posterior surface, the position of the antrum mastoid. (an) in relation to the posterior superior wall of the meatus, and on their anterior surface the relative position of the membrana tympani and the inner portion of the osseous meatus. They, moreover, enable us to obtain a clear view into the tympanum from behind, in which the topographical position of the ossicular chain, and the niches of the oval and round windows may be noticed.

Sections of dry preparations, in which the membrana tympani and

the ossicula are cut through, must be made with the finest saws, on
objects which have been hardened in alcohol, and afterwards
thoroughly dried. I possess a number of preparations of this class,
in some of which the malleus and incus, others in which the membrana tympani, the manubrium and the long process of the incus,
have been sawn through without the slightest dislocation of the small
bones. On wet preparations, or such as were saturated with glycerine,
the ossicles are nearly always dislocated in the sawing.

6. A horizontal section, on which the external surface of the
membrana tympani, the tympanic cavity with the ossicula, the
tendon of the tensor tympani, the vestibule, superior semicircular
canal (cs), the cochlea (co), and the internal meatus, come clearly
into view (Fig. 122).

The proceeding is as follows: First the tympanic membrane with
its osseous frame is laid bare in the way indicated sub 1. of this
section; then the tegmen tymp. is broken away with the forceps, and
the cavity of the labyrinth opened with a fine fret-saw in the follow-

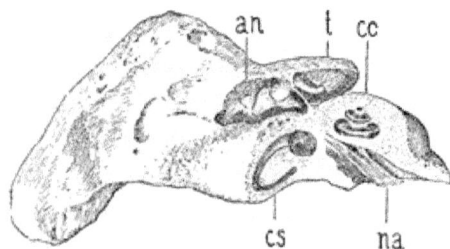

Fig. 122.—Topography of the tympanic cavity and of the labyrinth, after a preparation in
my collection.—t = membrana tymp. with the malleus and incus ; an = antr. mastoid. ;
co = cochlea ; cs = canalis semicircul. super., with the vestibulum ; na = nervus acusticus.

ing manner: the cut commences at a point vertically above the
eminentia arcuata, and takes a frontal direction corresponding to the
superior semicircular canal, advances downwards, with a very slight
curve forward, to the level of the internal meatus and the tendon of
the tensor tymp. Here it passes at nearly a right angle forward, to
divide the pyramid horizontally up to its apex. With some practice
it is possible, by these combined cuts, to represent the whole of the
superior semicircular canal (cs) with its ampullary dilatation, the
vestibule, the internal meatus with the nervus acusticus (na) and the
cochlea (co).

An instructive topographical preparation is obtained if, after laying
bare the membrana tympani and ossicula in the manner already described (sub 1 and 6), the pyramid be cut through in a perpendicular
direction corresponding to its long axis. If the cut be carried parallel

with the inner wall of the tympanic cavity through the middle of the vestibule, the lateral surface of the section presents the already described aspect of the external wall of the labyrinth, together with the inner surface of the footplate of the stapes, the commencing portion of the lamina spiralis, and, towards the front, the cochlea divided perpendicularly to its axis.

7. Very instructive for the topography of the organ of hearing are preparations on which, as the accompanying illustration (Fig. 123) shows, the membrana tympani, the ossicular chain with the tensor tendon, and the osseous labyrinth have been prepared out in one piece. The preparation of such objects is most difficult, as special manual dexterity and practice are required in order to cut out from

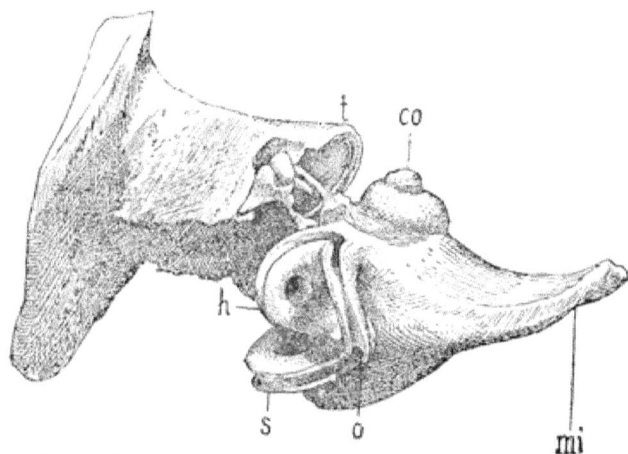

FIG. 123.—Topographical preparation, on which the tympanic cavity and the osseous labyrinth have been prepared out in connection. (Enlarged to double its size.)—t = tympanic membrane ; o = superior semicircular canal ; s = posterior semicircular canal ; h = horizontal semicircular canal ; co = cochlea ; mi = internal auditory canal. After a preparation in my collection.

the hard mass of the temporal bone of an adult the osseous semicircular canals and the capsule of the cochlea, without injuring at the same time the ossicles and the tympanic membrane. Working at the rate of two hours per day, it requires about a fortnight to complete one of these preparations.

A detailed description of the proceeding appears superfluous after what has been previously stated, since the preparation of the membrana tympani and tympanic cavity is the same as that described sub 5 and 6. This applies also to the preparation of the osseous labyrinth, respecting which we refer to the section under that heading (p. 115). The greatest difficulty is met with in laying bare the horizontal semicircular canal and the posterior portion of the cochlear convolutions, as the work on the exceedingly brittle bone has to be carried out with gravers and files, constantly moving backwards and forwards in close proximity to the tympanic membrane, which will be destroyed by the least slipping of the instruments.

8. Topographical preparations to show the inclination of the membrana tympani and the relative position of the Eustachian tubes.

On a skull from which the calvaria and the brain have been removed, the two pyramids are prepared out from the base of the skull in the manner described at p. 15, the tendon of the tensor tympani and the incudo-stapedial articulation having previously been divided from the upper tympanic space. The view of the inner surfaces of both tympanic membranes, now exposed on either side, gives a clear idea of their strong inclination to the horizontal. With the object of representing in the same preparation the position of the Eustachian tubes in the skull, and their relation to the tympanic cavity, those portions of the great wing of the sphenoid bone covering the cartilaginous Eustachian tubes, and the proc. pterygoid. are removed with chisel and forceps after the manner described on pp. 71 and 72, and the canal of the tube is then split longitudinally (p. 104).

Another method of demonstrating the position of the Eustachian tube in the skull consists in cutting the latter into halves in a sagittal direction, removing the inferior maxilla while leaving the septum narium intact; after which the tympanic membrane is laid bare by taking away the anterior lower wall of the meatus. The cartilagino-membranous tube is then prepared out of the surrounding parts in the following manner: A wooden cone, 3-4 cm. long, and running to a point, is pushed forward through the ost. pharyng. tubæ towards the narrowest part of the canal, in order, during the work of preparation, to allow the position of the latter to be recognised, and to avoid injuring its membranous portion. Leaving now the membranous tube in connection with the lateral choanal wall, all the adipose and muscular tissue lying round it, as also part of the basilar fibro-cartilage, are cut away with scalpel and scissors up to its union with the osseous tube, the whole tube being thereby exposed on all sides.

Mention might here be made of the mode of preparing some anatomical objects which are of importance for otoscopy.

1. A preparation, on which the osseous meatus in connection with the membrani tympani is prepared out of the temporal bone in the form of a tube. For this purpose the tympanic membrane is cut out with a fine fret-saw from the pars tympanica et squamosa on a moist or dry preparation which, after the removal of its soft parts, has been

thoroughly washed and then dehydrated by immersing it in alcohol. As the thin anterior wall of the meatus is merely bounded by soft structures, the saw-cut need only be carried round the upper posterior and lower wall of the meatus.   Such preparations serve for the superficial representation of the inclination of the membrana tympani to the axis of the osseous meatus, both for purposes of instruction, and for the study of the relationship of the different walls of the auditory canal to the tympanic membrane, which have to be considered when foreign bodies are to be removed from it.

2. An entire temporal bone, on which the soft parts are removed by maceration, the membrana tympani being retained intact. For this purpose, the fresh temporal bone, as it is taken from the body, according to the method recommended by Schwabach,* is placed in a

Fig. 124.—View, in projection, of the inner tympanic wall, showing its relation to the membrana tympani.—vo = anterior superior quadrant ; vu = anterior inferior quadrant ; ho = posterior superior quadrant ; hu = posterior inferior quadrant ; r = niche of the fenestra rotunda.

5-8% solution of caustic potash (Partsch's method of maceration), and the fluid, at first, renewed daily. In the ordinary temperature of a room the whole of the soft parts on the bone are in a short space of time destroyed by maceration, while the membrana tympani resists the action of the potash in the degree of concentration above mentioned, and remains intact on the macerated temporal bone. On specimens thus dried the membrane appears almost exactly as during life, and is, therefore, specially suitable for purposes of demonstration, previous to practising the examination of the membrane in the living subject.

3 A preparation to represent the topographical relationship of the membrana tympani to the inner wall of the tympanic cavity.

The process consists simply in cutting out the membrane, without at the same time dislocating the malleus and incus. With sufficient illumination from the reflecting mirror, the upper periphery of the membrane is first detached from its insertion by means of a sharp-pointed knife ; then two incisions are made in front of, and below the

* Centralblatt für medicinische Wissensch. 1885, Nr. 39.

manubrium down to its lower extremity, and finally a cut is carried round its entire periphery. Any portions of the membrane left behind are singed off with a red-hot piece of wire (Fig. 124). A glance into the tympanic cavity shows, in projection, the image of the inner wall, in its relation to the membrana tympani, acquaintance with which is of such importance, both in estimating the pathological appearance of the membrane, and when operating on this part, or in the tympanum. In order to obtain a general idea of the position of the various portions of the membrane in relation to the inner tympanic wall, the plane of the former is divided into four segments, by continuing the line of direction of the manubrium downwards, this line being crossed by another horizontal one, which touches the lower extremity of the handle.

Examining the membrane while the head is in the normal position we find, corresponding to the anterior superior quadrant (vo) of the membrane, the anterior superior portion of the inner tympanic wall bordering on the ostium tymp. tubæ (rarely only is part of the canalis pro tensore tymp. visible); corresponding to the anterior inferior quadrant (vu), the anterior inferior portion of the inner tympanic wall bordering on the ost. tymp. tubæ, and part of the ridged lower and anterior tympanic wall; corresponding to the posterior superior quadrant (ho), above, the incudo-stapedial articulation, and behind this the apex of the eminentia staped., with the tendon of the stapes; below that, the major part of the upper portion of the niche of the fenestra rotunda (r); and corresponding to the posterior inferior quadrant (hu), above, the minor inferior portion of the niche of the round window (r), and below, part of the ridged lower wall of the tympanum.

As regards the making of topographical sections on pathologico-anatomical preparations we must, in order to avoid repetition, refer to the methods of preparing such specimens of the organ of hearing described on p. 51 and following. It should, however, be observed that the practice, still in use in many anatomical institutions, of sawing the temporal bone through, in all cases of fatal middle ear suppurations without previous careful examination of the external and middle ear, is altogether objectionable, because, in doing so, not only are the soft structures destroyed, but the examination of the labyrinth, on account of its being filled with bone dust, is rendered impossible. Only in rare cases, particularly in necrosis of the pyramid, does it appear advisable to lay bare, by a corresponding cut with the saw, the breach in the bone leading from the temporal bone into the cranial cavity, whereby the extent of the disease in the interior of the petrous bone will be more clearly noticed. In caries and necrosis of the temporal bone a tolerably good idea, as to the extent of the morbid process, may be formed from macerated preparations, on which it is possible to recognise how far the bone is affected from its worm-eaten, eroded, and porous appearance, and from the osteophytic formations, of frequent occurrence in the neighbourhood of the seat of the disease. Maceration of the carious-necrotic temporal bone should, however, only be decided upon when the soft parts of the external meatus, the membrana tympani, the lining membrane of the middle ear, the labyrinth and dura mater have been minutely examined, and the changes found there present nothing requiring the preparation to be preserved in spirits of wine.

Though I have, on the grounds above mentioned, pronounced against the representation of topographical sections on pathological preparations, I must, on the other hand, lay stress on the importance of such sections in certain cases. These are the extensive adhesions between membrana tympani and inner tympanic wall, the connective-tissue proliferations enveloping the ossicula in the upper tympanic space, etc., the outlines and topographical position of which come out more clearly on a section than on an ordinary preparation. In these cases one ought, however, not to make use of the saw, which would destroy the soft parts, but the preparation should be decalcified after the method described on p. 156, and the section made with the knife.

V.

MAKING PREPARATIONS OF THE ORGAN OF HEARING BY CORROSION.

PREPARATIONS of the organ of hearing by corrosion form a material aid for the study of its anatomy. The three divisions being made up of cavities of more or less complicated shape, we obtain, by corrosion preparations, a far more precise notion of the actual form and extent of the external, middle, and internal ear than on sections carried in various directions through the macerated or unmacerated temporal bone. Of still greater importance is this method of preparation for the study of the topographical positions of the various parts of the organ of hearing, and of their relationship to one another. A good view as to the extent, especially of the pneumatic cavities of the middle ear, of their connection with one another, and their relationship to the external and inner ear and to the large neighbouring vessels, etc., can in general only be gained in this way. It is only by representing these spaces in the solid that the significance of their so frequently occurring inflammatory affections can be brought into the full light.

Historical.—The first attempts to produce preparations of the ear by corrosion were limited to the internal ear, and date from the second decennium of our century. According to Bezold, the first casts in wax of the osseous labyrinth appear to have been made by the veterinary surgeon Gerber at A. Meckel's suggestion.* Hyrtl (*Medicinische Jahrbücher*, 1873) saw in Meckel's museum at Halle and at Valentin's in Berne casts of the labyrinth. Also Ilg gives in his work on the structure of the cochlea an illustration of a metal cast of it. But it was Hyrtl who first employed the method of corrosion on a large scale for the study of the comparative anatomy of the osseous labyrinth of vertebrate animals, the results of which he has embodied in his celebrated work, copiously illustrated, ' Vergleichend-anatomische Untersuchungen über das innere Gehörorgan; Prag, 1845.' Rüdinger, too, in his atlas of the organ of hearing, has illustrations of casts of the labyrinth.   According to Sapolini (III. Otological Congress at Bale, 1884), there are in the Anatomical Museum of Turin very fine metallic casts of the labyrinth, the work of Professor Tommasi.

Interest for this method of preparation of the external and middle ear does not seem to have been awakened among anatomists until later. For although Sömmering in 1806, in his anatomical illustrations of the organ of hearing, shows a cast of the auricle and external meatus, the first preparations of the middle ear by corrosion are to be found in Hyrtl's work, ' Die Corrosionsanatomie und ihre Ergebnisse, 1873,' in which he gives a description, accompanied by several illustrations, of corrosion preparations of the middle ear.

To Bezold belongs the merit of having elaborated this method in all its directions, and of having, in his treatise ' Die Corrosions-anatomie des Ohres, mit 6 Tafeln in Lichtdruck, München, 1882,' utilized it for the study of anatomy and for practical otology. Referring, as regards the description of the anatomical details obtained through this method, to the excellent work alluded to, we here propose to discuss merely its technical side.

Preparations of the ear by corrosion are made either on macerated temporal bones (dry corrosion) or on fresh preparations (soft-part preparations).

For the purpose of anatomical study, both kinds of representation are indispensable. Soft-part corrosion preparations allow us to see at one glance the upper spaces of the middle ear, the cartilaginous and osseous Eustachian tube, tympanic cavity, aditus ad antrum, antrum, as well as the commencement of the pneumatic cells, and, by simultaneous representation of the external ear, the relative position of the one to the other. Preparations of the bone by

---

* Archiv für Anatomie und Physiologie von Meckel, Jahrgang 1827.

corrosion, on the other hand, bring into representation the entire system of pneumatic spaces in connection not only with the osseous meatus auditor. ext., but also with the labyrinth, the aquæducts, and the large vessels.

1. BONE-CORROSION PREPARATIONS.

To prepare the macerated temporal bone by corrosion, a thoroughly dried specimen, free from all remains of soft parts, should be chosen. In order to destroy any portions of tissue still remaining behind in the inner spaces of the temporal bone, and which often prove an obstacle to successful corrosion, the bone is left for several hours in a 5% solution of caustic potash heated to 60°; it is then washed in clean water and thoroughly dried, after a vigorous use of the syringe.

Before placing the temporal bone in the corrosion medium, a needle is pushed through the internal meatus, the vestibule, and fenestra ovalis, as far as the external meatus, for the purpose of keeping the various portions of the cavities in the temporal bone in close connection, two short needles being so inserted as to pass from the sulcus transversus through the mastoid cells. When the substance has hardened, these needles are cut level with it, but not removed, in order to preserve the connection between the cast of the sulcus transversus and that of the inner cavities of the temporal bone. To facilitate the escape of air from the labyrinthine cavity, it is recommended to make a small opening in the superior semicircular canal.

(a) Corrosion Preparations by the Medium of a Compound of Wax and Resin.

The perfectly-dried temporal bones are boiled in a melted mass consisting of four parts of resin and one part of wax, until the air has escaped from all the cavities. To uniformly colour the corrosion substance, either ultramarine, minium, or any other pigment is added, which has been previously triturated with a small quantity of copaiba balsam (Rüdinger). By the addition of the latter the mass also loses some of its brittleness.

The liquid having hardened, the wax covering the outer surface of the preparation is removed by scraping it with a heated knife, and the object placed in concentrated hydrochloric acid, in which the bone is more or less rapidly destroyed.

To retain the aquæductus vestibuli, a bridge of the medium is left between its orifice at the posterior wall of the pyramid (Fig. 105, p. 131) and the mass lying in the sulcus transversus. It is also advisable to leave a quantity of the corrosion medium adhering to the lower wall of the pyramid, which, during the work of preparation, serves as a firm and convenient handle.

Temporal bones of new-born infants, or of children during the first year of life, are corroded in 6-8 days, while bones of adults must be exposed for two or three weeks longer, to the action of the hydrochloric acid, to destroy any portion of bone adhering to the preparation.

When the maceration is completed, friable bits of the destroyed bone will be found in the depressions of the preparation. If an attempt be made to remove them with pincettes or needles, the more delicate parts of the specimen are broken off, whereby its scientific value is diminished. In order, therefore, to obtain an object as perfect as possible, the hydrochloric acid is carefully poured off, and

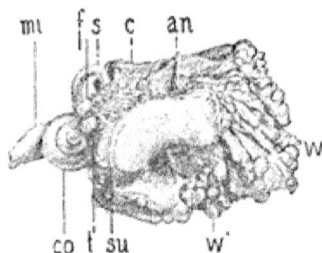

Fig. 125.—Corrosion preparation of the macerated temporal bone of an adult, made according to the wax and resin process; outer aspect, seen from the front. (Left ear.)—su = external meatus, with the sulcus tymp.; c = cavum tymp. and antr. mastoid.; w w′ = mastoid cells; t = cast of the canalis pro tens. tymp.; f = canalis facialis; s = canalis semicircul. super.; co = cochlea; mi = meat. audit. int. After a preparation in my collection.

the preparation lying at the bottom of the vessel is washed with the ball syringe, or with the more powerful stream of an ordinary syringe.

The moss-like impressions of the cavities of the spongy substance surrounding the corrosion cast on all sides must be removed with care. In the vicinity of the pneumatic spaces of the external meatus this is very easily done, but greater difficulty is encountered in the neighbourhood of the labyrinth and aquæducts, where some practice and intimate acquaintance with the relative positions of the parts concerned are needed. The spongy masses of the cast at the apex of the pyramid are so compact that they can, in most cases, only be separated with a narrow, sharp knife.

The cast of the macerated temporal bone (Fig. 125) when completed shows, in a more general manner than is possible with sections through the bone, the relations of the spaces of the osseous meatus in its various divisions, the length of its different walls from the external orifice to the sulcus tymp., the form of its transverse section, varying from without inwards, the relative curvature of its walls, as well as the direction of its course up to the membrana tympani. At the boundary between the meatus and middle ear the impression of the sulcus tymp. stands out at the lower and lateral periphery, in the form of a sharply-defined ledge. The cast of the tympanic cavity is, corresponding to its upper recess, found to project beyond that of the external meatus. From above, the cast appears at times smooth, at others, again, in accordance with the ridged condition of the upper tympanic wall, irregularly furrowed; one groove in particular, running in the transverse diameter of the tympanum, is to be regarded as constant, and corresponds, according to Bezold, to the constant transverse ledge on the upper tympanic wall (crista transversa tympani, see p. 29). Besides, there are found on some preparations of this kind, in great variety and to a large extent, the impressions of the pneumatic spaces, which enter into close relationship, externally,

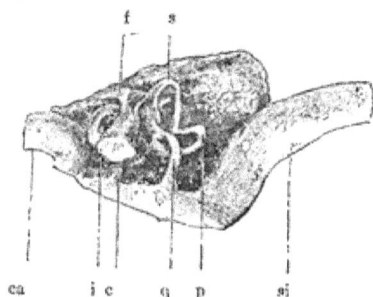

FIG. 126.—Wax and resin corrosion preparation of the temporal bone. View of the inner side, with the cast of the labyrinth.—i = meat. audit. intern.; s = canal. semicircul. super.; p = canal. semicirc. post.; q = aquædnct. vestibuli; c = aqueduct. cochleæ; f = canal. facialis; ca = canal. caroticus; si = sinus transversus. (After Bezold.)

with the upper wall of the outer meatus, internally, with the air spaces in the pyramid. On the inferior surface of the cast it is, from the position of the irregular cellular spaces of the lower tympanic wall, much easier to demonstrate than on frontal sections of the temporal bone, by how much the floor of the tympanic cavity is more deeply situated than the lower periphery of the membrana tympani, or the floor of the osseous meatus.

The upper and inner surface of a perfect cast (Fig. 126) bring under observation the three semicircular canals—the cochlea, the internal meatus (i), and the facial canal (f). The latter may, in correspondence with its anatomical position in the temporal bone, be traced with all its windings through the entire cast as far as the foramen stylo-mastoid. In addition is seen the aquæductus vestibuli (q), proceeding from the inner wall of the labyrinth as a fine threadlike loop, the lower, broader end of which corresponds to the hiatus aquæductus vestibuli on the posterior surface of the pyramid. More frequently still we see the cast of the aquæductus cochleæ (c), which, corresponding to the posterior lower edge of the pyramid below the commencing portion of the internal meatus, is connected, in the form of a short cone running upwards to a point, with the commencing part of the lowest convolution of the cochlea.

Objects prepared by the medium of the wax and resin compound must, on account of the brittleness and fragility of the corrosion substance, be handled and preserved with the greatest care. In order to render them more capable of resistance, Hyrtl recommends that the preparation be repeatedly coated with a solution of isinglass.

Experiments which I have lately made to cover preparations of

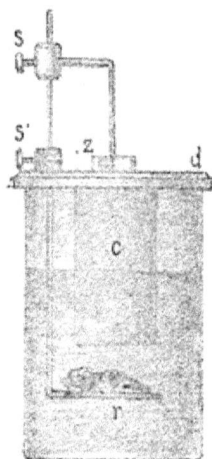

Fig. 127.

this kind with a galvano-plastic layer of copper have had satisfactory results.

I use for this purpose a small galvano-plastic apparatus, containing about 1 litre of liquid, such as is used by amateurs for making galvano-plastic copies of ancient coins. The apparatus (Fig. 127) consists of a glass jar closed by a wooden lid (d), and holding about $1\frac{1}{2}$ litre of fluid. Fixed to the lid is a porous pot (c), into which dips a grooved zinc rod, attached to a copper support. The holder of the copper support bearing the zinc rod is fixed in the wooden lid, passes through it into the glass vessel, and terminates in a round copper disc (r) with a bright upper surface. This, as well as the zinc rod in the porous pot, can be raised and lowered by adjusting-screws (s s'). The apparatus forms a galvanic element,

having the zinc rod for its negative, and the copper disc in the glass
for its positive pole. The outer vessel is filled up to two-thirds of
its height with distilled water, wherein 250 g. sulphate of iron and
50 g. bisulphate of potash are dissolved.

The glass jar having been filled, its lid, together with the copper
support and the zinc rod are put in position. Distilled water, with
10 g. of common salt dissolved in it, is poured into the porous pot
up to the level of the liquid in the glass, after which the apparatus
will be ready for immediate use.

In order to coat the cast with a galvanic layer of copper, its entire
surface, even to the least depression, must, by repeated brushing, be
first covered with a layer of finely triturated Siberian graphite. The
object is then placed upon the copper disc (r), and, by means of the
adjusting-screw (s'), brought within 1-2 cm. of the lower surface of
the porous pot. If, at the end of several hours, the surface of the

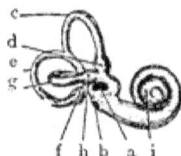

Fig. 128.--Cast of the osseous labyrinth.--a = fenestra ovalis ; b = fenestra rotunda ; c =
superior semicircular canal ; d = its ampulla ; e = posterior semicircular canal ; f = ampulla of the same ; g = horizontal semicircular canal ; h = its ampulla ; i = cochlea.

preparation, which is turned upwards, appear copper coloured, it is
turned, and left for several hours longer in the apparatus. When
coated with a deposit of copper, the preparation is washed in water
and cleaned with a soft brush. The thin layer of copper gives a
sufficient degree of firmness to the preparation without materially
altering the form. To prevent the copper oxidising, such casts may,
at small cost, be silvered or gilt.

For the corrosion of the labyrinth alone, by the wax-resin process,
well-macerated pyramids from temporal bones of adults are the most
suitable. It is more difficult to obtain casts of the labyrinth from
petrous bones of new-born infants, because here the corrosion medium
penetrates into the porous bone up to the immediate vicinity of the
capsule, and the cast can only with difficulty be separated from the
net-like masses which surround it on all sides. The filling of the
labyrinthine cavity with the corrosion mass is effected, either by
boiling in the wax and resin, or by injecting it with a heated glass
tube. The cast is the more successful if the capsule be so far cleared
from the surrounding bone, that at the highest point of convexity of
each semicircular canal, and at the anterior edge of the lower convo-

lution of the cochlea, an aperture of the size of a pin's head can be made, which allows the escape of air from the labyrinthine cavity, and permits the wax to completely fill it. Casts of the labyrinth may also be made in a space where the air has been rarefied, a method which is to be discussed more in detail in the description of metal corrosion apparatuses.

Should the preparation, after setting of the corrosion medium, be sufficiently warmed as to allow part of it to flow again from the labyrinthine cavity, a cast is obtained after corrosion which, as Hyrtl asserts, may be mistaken for the natural osseous labyrinth.

(b) *Representation of Metallic Corrosion Preparations of the Organ of Hearing.*

Metallic corrosion preparations are, on account of their durability, eminently adapted for the purposes of instruction. The alloy when melted reaches the minutest canals of the petrous bone as readily as the wax and resin compound does. But while, with the latter, the casts of the finest channels break, even during corrosion, they are preserved perfect in the metallic casts.*

For making metallic corrosion preparations, various kinds of alloys are made use of, which are fusible at a relatively low temperature. Most suitable are found to be : 1st, Métal d'Arcet, consisting of 1 part lead, 2 parts tin, and 2 parts bismuth ; 2nd, Rose's Metal, consisting of 2 parts bismuth, 1 part lead, and 1 part tin ; 3rd, Wood's Alloy, consisting of 8 parts lead, 4 parts tin, 16 parts bismuth, and 3 parts cadmium. The latter is adapted, on account of its ready fusibility, especially for casts of the labyrinth.

The proceeding in making metallic casts of the temporal bone differs materially from that observed with wax-resin corrosions. Plunging the temporal bone into the molten alloy nearly always results in imperfect casts. Simply pouring the alloy into the external meatus does not, however, yield the desired result, because the metal escapes through the numerous openings in the temporal bone, which are in communication with each other.

For the production of metallic casts of the temporal bone, special methods are therefore recommended, a brief description of which we give in the following :

Professor v. Brunn, of Rostock, showed, at a meeting of German naturalists and medical men held in Berlin in 1886, a series of successful metallic casts of the temporal bone, which were produced by the following method, for the communication of which we are indebted to Dr. Lemcke :

---

* Passage altered at the request of the author.—G. S.

To make perfect metallic casts of the external, middle, and internal ear on macerated temporal bones, the external meatus is closed by a cork, the porus acust. int. pasted over with a small piece of paper, and then the entire temporal bone covered with gypsum, so that only the opening of the osseous tube remains free. The whole is then placed in the heating apparatus, and left there for three or four hours at a temperature of about 100° C. After this the metal is poured in through the opening, and by repeatedly shaking the object, made to penetrate into every cavity. Generally all the spaces of the temporal bone are completely filled.

The preparation having cooled, the plaster is removed, the bone transferred into 8% caustic potash, and the vessel placed in the heating apparatus (50° C.). In a few days any remains of bone are washed away by a stream of water, and the cast will be found perfect.

At the Otological Congress at Brussels, Dr. Siebenmann demonstrated a series of perfect metallic casts of the organ of hearing made after a method of his own, which in several respects differs materially from that of Professor v. Brunn. We here propose to give a short description of this method.

To make casts of macerated temporal bones, it is, on the whole, better to select such as are not too heavy, and have been carefully macerated. Before the casting is proceeded with, the following preliminary preparations have to be attended to : the canaliculi carotico-tympanici are widened with a needle, and from the canaliculus caroticus an opening is made in the tube. The superior semicircular canal is slightly opened with a file behind the crista. The sulcus transversus and, in connection with it, the base of the petrous bone, are spread over with strips of linen in order that they may subsequently be covered with a thicker layer of metal. In the same manner a canal bridged over is to be made from the external opening of the aquæduct. vestibuli down to the sinus, so as to give to the cast of the canalis aquæd. a firmer support, and the openings of the meatus extern. et intern. must be closed with linen. On the apex of the pyramid a cardboard funnel, not too narrow at its lower end, is introduced into the canalis caroticus and likewise firmly glued on with paper or linen strips.

When the casting is to commence, the bone thus prepared should first be well warmed ; it is, moreover, indispensable that the object should be firmly covered on all sides by a material which, though giving free passage to the air, does not allow the metal to escape. This was done, in the production of the specimens demonstrated at Brussels, by heating the bone in a covered pan, quickly wrapping it first in linen cloth, then in wadding, and finally surrounding it, with the exception of the funnel, on all sides with moulding clay to the depth of 5 cm. By vigorous kneading with both hands this was shaped into a firm ball.

Extremely good casts are obtained by the following method, which has of late been universally employed : the bone, provided with linen bridges, as indicated above, is covered with plaster of Paris as far as the edge of the funnel, which is turned upwards, and then left to stand for several days in a moderate temperature with the funnel turned downwards. The filling of the block should not be proceeded with until it is thoroughly dry, and while allowing the air to escape freely, it should not show any cracks in the heating, which must necessarily precede the casting.

If the object has been well cleaned and sufficiently heated, metal poured in through the funnel will fill up any, even the minutest cavity from the squama to the apex of the pyramid. In the preparations which were made last, not only the canals of the two aquæducts, but also those of the veins accompanying them, were represented in the most perfect manner.

While pouring the molten metal in a continuous stream, and in as large a quantity as possible, through the funnel, it is well to gently knock the bottom of the block of plaster against the table. The pouring in of the metal is only to cease when its level no longer sinks in the funnel; after waiting a few minutes, the block is placed upright in water. On the following day the plaster is broken away, the bone cleaned with a brush, and the metal removed with the chisel or a red-hot blade of a knife, on those spots where it is not required.

The petrous bones are next corroded in 10-30% caustic potash (or caustic soda), at a temperature of 40-50° C. At the end of a month they are soaked for 24 hours in water, and then placed in dilute (1 : 6) pure hydrochloric acid. After a few hours the last remains of the adhering portions of bone (labyrinthine capsule, etc.) are dissolved, so that nothing further remains to be done but to free the preparation from the acid by placing it in distilled water. During the entire process of corroding, the preparation never comes under a stream of water, which is rather an advantage over other corroding methods in use.

For further preservation, and for purposes of demonstration, the specimen is fixed on a forked stand with heavy supports. The branches of the fork are, with this object, forced when red-hot into the thick metal plate on the lower surface of the preparation in such a way, that, when fixed, it has as nearly as possible the natural position of the petrous bone in the upright skull.

In order to make such casts more serviceable for study, it is advisable to brush over the various parts with a thin layer of paint (or varnish), *e.g.*, the meatus bone-yellow, the nerve canals bright yellow, middle ear bright rose-colour, labyrinth greenish-yellow, the carotid artery crimson, sinus bulbus jugul. and veins of the aqueduct light blue, diploë brownish. On preparations which are intended to give a general view of the organ the spongiosa is first broken away. Preparations which are not to be painted over are lacquered with 'Fixative.'

The proceeding in the making of metallic corrosions of soft-part preparations (casts of the middle ear), after Siebenmann's method, is identical with that of Bezold (making use of the wax and resin compound). For the purpose of more thoroughly heating the moist temporal bones, and of more surely fixing the cartilaginous tube, Siebenmann also uses gypsum ; only the opening of the tube and the artificial gap in the proc. mastoid. are left free. But the plaster block, having been hardened, is immediately heated, and the casting proceeded with. The syringe to be employed must be one of steel or brass, and should be well heated before using it. The air-tight closure between the syringe and the gap in the mastoid process is simply effected by fastening on the nozzle of the syringe a short piece of drainage tube with thick walls but of small calibre, in such a way that the nozzle projects only about 3 mm. beyond it. The selection of suitable bones, the further preparatory arrangements, and the injection are the same as in Bezold's process. For corrosion, however, the well-cleaned bone is merely placed in 20% caustic potash, where, especially under temperatures of 40-50° C., the destruction of the temporal bone is so rapidly effected, that in two, at most three or four, weeks the bright metallic preparation can be removed from the lye. For the corrosion of the moist temporal bones of children, only a few days are required.

Instructive preparations may be obtained by immersing such bones when charged with metal for a few days in a very weak solution of the caustic, until the soft parts and the diploë can be easily removed. At this stage the bones are again soaked for a day or two in water and cleared of the soft parts, care being taken not to injure the tube. Now the outer osseous plate of the lateral surface of the petrous bone is removed ; the air-spaces of the anterior and posterior surfaces of the pyramid must be carefully exposed and freed from the surrounding spongy tissue ; the outer anterior wall of the tubal canal is broken away in its entire length, and in some preparations also the inner half of the anterior lower wall of the meatus. If the bone has been corroded too far, it is, after the filling of the middle ear has been chiselled away or dug out, to be dipped once more into thin

12

liquid glue or gelatine, to restore its firmness, and to prevent the metal falling out.

Several good metallic corrosions of the temporal bone I obtained by the following simple procedure: on a well-dried, markedly pneumatic temporal bone, a small hole is bored $\frac{1}{2}$-1 cm. behind the external orifice of the ear. After this the anterior opening of the osseous Eustachian tube, and the hiatus canalis Fallopii are stopped up with glazier's putty, and the preparation is embedded in heated dry sand, so that only the external opening of the osseous meatus and the drill hole in the mastoid process turned upwards, remain free from the sand. A small quantity of the molten alloy is now poured into the auditory canal by means of a pointed spoon, in such a way that the metal penetrates through the round and oval windows into the cavity of the labyrinth; the tympanic cavity and the external meatus are then filled, until the alloy runs out from the bore-hole in the mastoid process. Experiments which I have lately made to produce metallic corrosions by injecting the alloy through the internal meatus have yielded a favourable result. With this object the osseous lamella, separating the floor of the internal meatus from the cavity of the labyrinth, is pierced with a pointed instrument, and another hole made on the processus mastoid. behind the external meatus. Then the preparation is covered over with gypsum, so as to leave only the external and internal meatus, and the bore-hole in the mastoid free. The injection of the alloy from below upwards is effected by means of a heated metal syringe, the nozzle of which, covered with a piece of india-rubber tube, fits air-tight into the internal meatus. The injection is continued until the mass flows from the bore-hole in the mastoid process.

Metallic casts of the labyrinth may be made simply by injecting the alloy through the fenestra ovalis, or by making the injection in a space where the air has been rarefied according to Gottfried's method. This proceeding, of the excellence of which I had the opportunity of convincing myself from the perfectly successful preparations of the labyrinth made by Dr. Dalla Rosa, consists in placing the pyramid, reduced down to the capsule of the labyrinth, together with several small pieces of Wood's metal, in a spacious éprouvette. The latter is stopped by means of an india-rubber plug with a hole in it, through which a short glass tube is passed. Now, in order to rarefy, as much as possible, the air in the éprouvette, and to facilitate the penetrating of the molten metal into the labyrinthine cavity, water is made to boil in a retort, provided with a forked tube, over the flame of a spirit-lamp, until the steam escapes with some force at the openings of the tube. If then the mouth of one of the branches be connected by means of an india-rubber tube with that of the éprouvette, the steam can only escape by the other branch, whereby the air in the éprouvette becomes so rarefied that the metal, which is melted at the same time over a spirit-lamp, penetrates into all the canals of the labyrinth.

The maceration of the objects with caustic potash requires the same precautions to be taken as with Siebenmann's method.

2. CORROSION PREPARATIONS OF SOFT PARTS.

On corrosion preparations of the soft parts of the temporal bone, only the air-containing spaces of the external and middle ear can be represented. The procedure differs, according to whether a cast of the auricle and external meatus, or one of the middle ear (tympanic cavity, mastoid cells, and Eustachian tube), or finally, one of the external and middle ear in connection, is to be obtained.

The easiest to represent are casts which show the depressions of the auricle, the length and direction of the external meatus, and the curvature of the outer surface of the membrana tympani. For this purpose a normal organ is made use of, the external meatus of which has been thoroughly cleansed from ceruminal and epidermic masses by syringing, and where the tympanic membrane, upon examination with a mirror, is found to be in a perfectly normal condition. Before pouring in the corrosion mass, the preparation must be warmed in hot water, and after it has been taken out, the meatus should be well dried by repeatedly forcing air into it from a heated metal syringe. As a corrosion mass, either that of wax and resin, before described, or one of the alloys mentioned, may be used (Löwenberg). These latter, according to Professor Brunn, should not be poured in too hot, otherwise the cutis of the external meatus, and the membrana tympani might be burned or shrunk, and the casts, in consequence, turn out shapeless and thin. When pouring in the metal, special care should be taken that the meatus and auricle be kept in their normal positions, lest by traction on the latter, the relationship of one to the other might be altered. When using metallic alloys, the angular incurvations of the meatus stand out in too pronounced a manner in the cast, owing to the weight of the metal filling the auricle, while the impression of the membrana tympani is less distinct.

After the wax and resin compound has hardened, the preparation is placed in concentrated hydrochloric acid, and the corrosion proceeded with in the same manner as with macerated temporal bones. With metallic casts, after cooling of the molten alloy, the auricle and cartilaginous meatus are cut away, and, according to Prof. Brunn, the cast may then be taken out. If this cannot be done the pars

tympanica must be removed with the chisel, or destroyed by 8%
caustic potash.

Somewhat more difficult is the procedure with corrosions of the
middle ear. The object selected must be taken from the skull in such
a manner that, towards the front, the Eustachian tube remains intact,
while the posterior cut through the bone is so directed that the cells
of the mastoid process are not opened thereby. The injection into the
spaces of the middle ear is best effected, according to Bezold, from
an opening made in the incisura mastoid., since, especially in pneumatic
temporal bones, there frequently exist, in the neighbourhood of this
incisura, one or several large cavities which communicate with the
mastoid antrum. Having introduced the canula air-tight into this
opening, warm air is made to pass through the middle ear by means

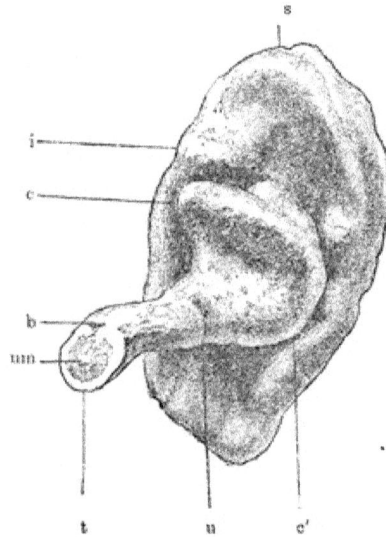

Fig. 129.—Corrosion cast of the auricle and of the external auditory canal. (After Bezold.)
—s = fossa scaphoidea ; i = fossa intercruralis, c = upper, c′ = lower portion of the
concha ; u = second bend of the external meatus ; t = border of the tympanic membrane ;
um = umbo ; b = depression of the membr. Shrapnelli, and of the proc. brevis.

of a large, well-heated metal syringe. If, while this is being done,
air escapes from the ostium pharyng. tubæ, it may be assumed that
the cavities in the mastoid process communicate with the rest of the
spaces in the middle ear.

Previous to injecting the corrosion mass into the middle ear, an
insect pin is pushed through the tubal canal into the tympanic
cavity, so that when the injection is completed, the pin will be surrounded by the corrosion substance, and the breaking of the cast of

the tube at the isthmus tubæ will be prevented. The heated corrosion medium is now injected with some force through the canula at the incisura mastoid., until it flows out freely, and without air-bubbles, from the ostium pharyng. tubæ which is directed upwards. The preparation is kept in this position until the corrosion mass has hardened, and must afterwards be immersed, for the purpose of corroding it, in concentrated hydrochloric acid (where alloys are used, in caustic potash). If, on corrosion preparations of soft parts, the external and middle ear are to be represented in connection, several strong insect pins are pushed from the meatus through the membrana tympani. The injection of the spaces of the middle ear is effected after the method previously described. When the corrosion medium has cooled in the spaces of the middle ear, the preparation is so placed, that the external meatus is directed vertically upwards, and then the latter and the auricle are likewise filled with the liquid mass. The

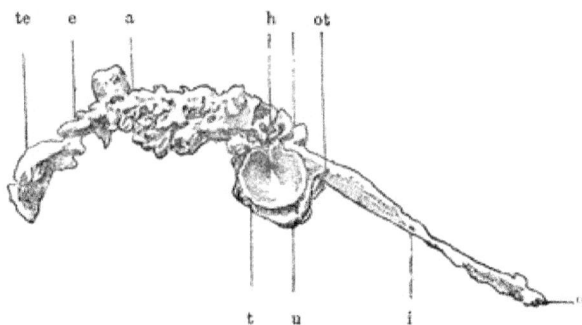

FIG. 130.—Corrosion cast of the middle ear. (After Bezold.)—o = ost. pharyng. tubæ ; i = isthmus tubæ ; ct = ost. tymp. tubæ ; u = lower tympanic space ; t = tympanic membrane, with the groove of the malleus, and the umbo ; h = depression for the body of the malleus and incus ; a = posterior end of the mastoid antrum ; e = transition cell : te = terminal cell.

connection between the spaces of the external and middle ear is preserved by means of pins thrust through the tympanic membrane.

The cast of the soft parts of the middle ear differs in many respects from the middle-ear cast of the macerated temporal bone. With the former we obtain an exact impression of the inner surface of the tympanic membrane, which on bone corrosions must of course be wanting. On soft-part corrosions we also find, besides the hollow spaces, which have been left behind after the corrosion of the malleus and incus and of the stapes, the canal formed by the tendon of the tensor tympani, passing transversely through the cast of the tympanum, as well as the furrows which the larger mucous membrane folds leave in the corrosion mass. All these details are missing in bone corrosion, though we obtain by it the impression of the facial canal, while in the cast of the soft parts this impression, as well as those of the small osseous canals of the temporal bone, are wanting. In the mastoid process, only the pneumatic spaces appear on the soft-part cast; on the cast of the macerated bone we have also those of the diploëtic spaces. Finally, we obtain on the macerated temporal bone only the cast of the osseous Eustachian tube, on the cast of the soft parts the entire tube. The cast of the latter does not correspond to the actual capacity of the tubal canal in a state of rest, but during its maximal dilatation.

Professor Steinbrügge recommends, for corrosion preparations of the labyrinth, celloidin, which of late has been so much used for embedding microscopical preparations.*

He opens the tympanic cavity from the roof, divides the musc. tensor tymp. and the incudo-stapedial articulation, and then, sawing through the antrum mastoid., separates the labyrinthine portion from the outer wall of the middle ear. The stapes is next carefully removed from the fenestra ovalis, and the bone then, for forty-eight hours, immersed in equal parts of ether and absolute alcohol, and afterwards placed in a thin-fluid solution of celloidin, which, after having been left for eight days in action, is allowed to evaporate by removing the lid of the vessel. The bones embedded in celloidin are left for fully three days in 50% alcohol; they are then freed from the superfluous celloidin by scraping, and placed in pure hydrochloric acid. After the lapse of three days the macerated tissues can be removed from the casts by means of a gentle stream of water. In this manner Steinbrügge obtained very good preparations of the perilymphatic spaces of the labyrinth, including the membranous structures; but his attempts to represent the latter in their connection by dissolving the celloidin again have hitherto not been successful.

Another method of obtaining preparations by the process of corrosion was suggested by Barth.†

In his various experiments to embed the ear he hit upon a plan of making casts of the spaces of the labyrinth, retaining the soft parts. The undecalcified and unopened petrous bone, after being hardened and stained (osmic acid), is placed first in dilute, then in absolute alcohol, and finally in chloroform. In each of these it is allowed to remain for several days. After that, paraffin is melted at as low a temperature as possible, the preparation quickly taken direct from the chloroform into the melted paraffin, and left there until the chloroform ceases to give off vapours, which is, with a human petrous bone, from five to six hours. After allowing the paraffin to cool rapidly, the

---

* Zur Corrosionsanatomie des Ohres. (Centralbl. f. d. med. Wissenschaft, 1885, Nr. 31).
† Beiträge zur Anatomie des Ohres, Z. f. O., Bd. XVII. s. 261-266.

preparation is cut out, a small block of paraffin being left standing for subsequent attachment to the meatus auditorius internus, or to the fenestra ovalis et rotunda. In the latter case the stapes must, of course, be removed before immersion in the medium, and the membrana tympani secundaria pierced. The superfluous paraffin is scraped off down to the bone. Now the petrous bone is transferred to concentrated or nearly concentrated common hydrochloric acid, in which it remains, according to its bulk, for eight to fourteen days. The preparation, on which a considerable portion of the cast of the inner ear is already visible, is then to be carefully placed in water, and what remains of the tissue still adhering to it is washed off by means of a syringe provided with a long nozzle.

Thus a cast of the whole of the internal ear is obtained, which includes the soft parts, sometimes with, sometimes without, preservation of the periosteum of the scalæ. It shows everything that is to be noticed on a cast of the osseous labyrinth, and in addition, though only faintly shining through, the position of the n. facialis in its relationship to the acusticus, the origin of the nerve-trunks and their ramifications on the ampullæ, the radiation of the r. cochleæ, the course of the ductus cochlearis, as well as that of the membranous semicircular canals within the osseous. With organs of children, up to the age of about ten years, the internal ear is obtained in tolerably close connection with the nerve-trunks. The preparation being completed, may, for preservation, find a suitable place beside the cast of the internal meatus. In adults the nerve-passages to the internal ear have, through gradual ossification, become so narrow that the preparation must be very carefully handled to avoid breaking the cast. Barth's attempts to clear the preparation in the paraffin embedding— i.e., to render it quite transparent—have hitherto not been successful; but it gains somewhat in firmness, even against moderate heat, if it be placed for some time in soluble glass and dried, any defective places being subsequently brushed over with the same substance. In further experiments it is recommended to try careful painting over with resin dissolved in chloroform. For microscopic examination these preparations are unsuitable, although the pillars of Corti's organ, as well as cells and nuclei, are for the most part plainly recognisable; but the membranous labyrinth may be demonstrated by dissolving and removing the paraffin. This manipulation—dissolving in chloroform, gradual removal into ether, alcohol, glycerine—must be very carefully carried out, lest there occur disfiguring shrinkings and displacements, or even rents in the vestibular structures, or, what is more likely still, in the canals. Tolerably good preparations may, however, be obtained, which afterwards can be stained, to make them stand out more distinctly in glycerine. Barth considers this method better suited for the examination of other cavities. Celloidin casts become more resistant if the pyramid be placed in oil, or in resin dissolved in alcohol.

VI.

MOUNTING AND PRESERVING ANATOMICAL AND PATHOLOGICAL PREPARATIONS OF THE ORGAN OF HEARING.

PREPARATORY METHODS.

1. *Soaking the Preparations in Water.*

ALL preparations intended for preservation in an anatomical collection must, in a fresh state, either before or after preparing them, be soaked in clean water, which is frequently to be changed so as to get rid of the blood which has been left behind in the bone and soft parts. The space of time within which the blood may be removed from the temporal bone varies from one to three days. It is effected the more rapidly the oftener the water is changed, and frequently one day will be sufficient if the object be placed in a stream of water. The latter is advisable to prevent rapid putrefaction setting in, especially during the warmer season of the year.

When the object is quite freed from blood—which will be seen if, after repeatedly changing the water, the latter no longer shows any reddish tint—it is, for the purpose of dehydration, steeped in ordinary spirit, from which it may be taken at any time when required for preparation. This, however, is only to be recommended in cases where normal organs of hearing are to be dissected. Pathological objects, on the other hand, should always be prepared in the fresh state, as the new formations of tissue, the result of the morbid process in the ear, undergo alterations when soaked in water. If intended for preservation as macroscopic objects, they must also be steeped in water after their preparation is completed, and a note made of their appearances; they are then placed in spirits of wine. Alcohol preparations, if not quite free from blood, show the details less sharply.

2. *Bleaching of Preparations of the Ear intended for Preservation.*

Macerated temporal bones are, under favourable conditions, best bleached in the sun. Frequently, however, one meets with temporal bones which contain such quantities of fat that, even after long exposure to the rays of the sun, they do not lose their yellow colour. The best bleaching substance for such objects is benzoline. The manipulation requires, however, a special apparatus. This consists in a correspondingly large, hermetically closed metallic case in which, close to the bottom, a piece of wire netting is fixed for the temporal bones to rest on. If, now, a quantity of benzoline be poured on the floor of the case, the latter being well closed, the fat is extracted from the bone by the vapours arising, and the objects are completely bleached.

Another simpler method of bleaching yellow bones is to immerse them in a 10% aqueous solution of ammonia; the bones must, however, be exposed for several weeks to the action of the liquid, which should be frequently changed. The bleaching of temporal bones in chlorine water or in a solution of chlorinated lime is certainly more expeditiously accomplished; but the osseous tissue, particularly in young individuals, is too strongly affected by the destructive action of the chlorine and the hydrochloric acid. Smaller bones, such as the osseous labyrinth, the tympanic membrane with its osseous frame, and the ossicles, may be bleached in as dark a place as possible, either in a mixture of sulphuric ether and alcohol, or in a 6% solution of peroxide of hydrogen, with the addition of a few drops of ammonia.

With regard to the bleaching of unmacerated preparations of the ear, in the majority of cases thorough soaking in water and subsequent immersion in alcohol will suffice to whiten the bone. Only with exceptionally yellow or yellowish-brown bones may a bleaching medium be used. One of the best modes is to immerse the preparation in water slightly acidulated with nitric acid ($\frac{1}{2}$-1,000).

MOUNTING OF THE PREPARATIONS.

In mounting anatomical and pathologico-anatomical preparations of the auditory organ regard should be had, in the first instance, that the important details on each object be clearly and fully brought under observation. There is a distinction in the mode of mounting spirit preparations and dry preparations.

1. *Mounting of Spirit Preparations.*

Preparations preserved in spirit must, for the study of the anatomical details, almost invariably be taken out of the vessel where they are kept. It is, however, advisable, when placing the object in spirit, to give it such a position that the principal detail may be easily seen through the glass without there being any need of opening the vessel. The glass intended for keeping the preparation should, therefore, be only of a diameter corresponding to the size of the object, to minimize the changes of its position, which would interfere with the proper examination.   Such changes frequently happen when the vessels are too wide.

The greater number of the anatomical and pathologico-anatomical preparations in my collection are preserved in glass jars, provided with well-fitting glass stoppers; they measure 9 cm. in height, 6 cm. in width, and 4 cm. at the mouth.   All these are preparations on which the opened osseous meatus, the outer surface of the membrana tympani, the tympanic cavity and the pyramid in connection, are represented in sections of the most varied kinds.   Preparations on which, in addition to the above-mentioned parts of the ear, the auricle, mastoid process, and the Eustachian tube are retained, must be placed in larger vessels, having a height of 12 cm. and a width of 7 cm.

More bulky preparations, such as half the skull with the organ of hearing ready prepared, both organs in connection, frontal sections through the entire skull and ear, must, of course, be put into correspondingly capacious glasses.   Best adapted for these are longish oval shallow ones, 16 cm. long, 7 cm. wide, and 18 cm. high, which can be closed by a well-fitting glass lid (by means of glazier's putty).

Flat sections of the entire organ, particularly horizontal and frontal ones of decalcified preparations, on which the details appear more sharply marked in the spirit than when taken out of the liquid for inspection, are kept in commoner glasses, measuring $6\frac{1}{2}$ cm. in height, and 5 cm in width; these may be covered with a glass disc and hermetically closed with glazier's putty.

Thin sections of decalcified organs which, on account of their pliancy, curl up in the glass are fixed on uncoloured or coloured (blue, yellow, black) glass slips of suitable size.   The most durable fixing material, which is not soluble in alcohol, is ichthyocolla or dissolved glue gelatine.   A few drops of the strongly condensed solution, applied to the upper and lower edges of the preparation, which for a short time previously must have been lying in water, will be sufficient to fix it on the slip, and if the whole be carefully placed in spirit, the preparation will remain fixed in its position.

It would lead too far to enumerate all the preparations and sections of the organ of hearing it might be possible to make, and which might be found suitable for preservation in spirit for instruction purposes.   Besides, this appears to me superfluous, inasmuch as, from the description of the anatomical preparations hitherto given, it may be readily gathered which of them are required for study and demonstration.   It need here only be pointed out in a general way that for such purposes spirit preparations and dry preparations supplement one another, as some details appear more strongly marked in the former than in the latter. To demonstrate the soft parts, in particular, of the auditory apparatus, it is advisable to keep in every anatomical collection a considerable number of spirit prepara-tions, on which should be represented, in their mutual relationship: the auricle, the cartilaginous meatus, the tympanic membrane with its pouches, and the chorda tymp., the lining membrane and ligaments of the middle ear, the intra-tympanic muscles, the Eustachian tube with its muscular apparatus, the membranous labyrinth and the acoustic and facial nerves. For pathologico-anatomical preparations preserving in spirit is, in the majority of cases, preferable to dry mounting. Certain changes, such as opacities of the tympanic membrane, non-adherent cicatrices, and atrophic depressions, almost completely disappear when the objects are dried. On the other hand, calcareous deposits on the membrana tympani appear more sharply defined in dry than in spirit preparations; this applies also to adhesion bridges between the membrane, inner tympanic wall, and the ossicula.

Smaller preparations, intended for demonstration, are best kept in small, specially manufactured glasses such are shown in their actual size on p. 191, Figs. IV. and VI. Among these preparations are to be mentioned: transverse sections of the cartilagino-membranous meatus, the membrana tympani, with the annulus tymp.; the membrane, prepared out of the sulcus and dissected into its various layers; sections through the longitudinal axis of the decalcified malleus and of the membrana tympani, with the view of Shrapnell's membrane and of Prussak's space in profile; longitudinal, and trans-verse sections of the isolated Eustachian tube, the malleus, with the m. tens. tymp. dissected out; the stapes, with the m. stapedius, that portion of the inner tympanic wall on which the stapes with the m. staped., the fenestra rotunda and the n. facialis have been pre-pared out; likewise the external wall of the labyrinth, with the foot-plate of the stapes, and the commencing portion of the lamina spiralis, the sacculus with the membranous semicircular canals, the membranous cochlea, sections of hardened and decalcified petrous bones, etc. Especially clear and beautiful appear stained sections of the tympanic cavity, with the transverse section of the malleo-incudal articulation, of the longitudinal section of the malleus, with the view of Shrapnell's membrane and Prussak's space in profile, of the cochlea and vestibule, with the transverse sections of the saccules and ampullæ. Microscopical sections of those structures which have turned out too thick for histological examination form, when pre-served in small glasses, and examined with a lens in a good light, very beautiful and instructive objects for lecturing purposes. In order to mount such sections on an even surface, they are fastened at both ends by means of a drop of fluid gelatine to a small coloured (yellow or blue) glass slip, and after a few minutes placed in a small glass containing common spirit.

Of pathological preparations to be preserved in smaller glass vessels we may mention : calcareous deposits, perforation, cicatrices, granulations and polypoid formations on the tympanic membrane,[*] polypi, carious ossicles, the labyrinthine capsule wholly or partially exfoliated by necrosis, the extruded cochlea, etc.

For a preservative fluid I use the ordinary 50% spirit, in exceptional cases, for small objects (membranous cochlea, semicircular canals, membrana tympani, etc.), Laskowski's carbolized glycerine (1-5%).

It is of great importance, when forming an anatomical collection, to have well-fitting covers for the preparation glasses, because many objects, after the evaporation of the alcohol, dry up so as to become useless. Glasses closed with imperfectly fitting stoppers, and which, on being turned upside down, allow the liquid to escape, if only in drops, must be rejected and replaced by others. To ensure air-tight closure, the circumference of the glass stoppers is brushed over with a mixture of stearine and ceresine. The evaporation of the preservative fluid is best prevented by making use of glazier's putty, which can, however, only be done (see p. 186) with preparations which do not always require to be taken out of the vessel for demonstration. Preparations which, for want of attention, dry up through evaporation of the alcohol, are made to expand again by placing them for one or two days in water, strongly acidulated with nitric acid (2-3%), after which they are replaced in alcohol. When the preservative fluid is found to have turned brown or thick, it must of course be renewed.

2. *Mounting of Dry Preparations.*

The mounting and arranging of dry preparations is of the greatest importance if they are to be kept in a perfect condition. Special care should be taken in fixing them, that those portions of the object which are to be brought under notice should present themselves freely and clearly to view. Preparations of the tympanic cavity and labyrinth, in particular, must be mounted in such a way that their respective cavities be open to unobstructed inspection. It is scarcely necessary to say that, while mounting preparations of the auditory organ in a manner most suitable for their particular purpose,

---

[*] Compare the figure in my text-book of Diseases of the Ear, second edition, p. 188.

regard should also be had to their tasteful grouping, in order to show them to greater advantage.

Smaller preparations, such as the outer and inner surfaces of the membrana tympani stretched out in the sulc. tymp., the ossicles, sections through the vestibule and cochlea, are best fixed with a drop of glue on small black slips of wood measuring 4 cm. by 2 cm. Glass slips (reduced to the proper size) have the advantage of allowing the objects fixed upon them to be examined on both sides. In order to protect the preparation from dust and insects, the slip is fastened on a round stand, which has a diameter of 4 cm., and is covered with a glass shade 5 cm. (high p. 191, Figs. I. and III. in the actual size).

In a like manner, smaller objects may be mounted in a round capsule resting on a stand, and closed in front with a watch-glass fitted in (p. 191, Fig. II.). A still more simple method of mounting consists, as shown in Fig. VII., in fixing the preparation with a drop of glue on a small, black polished or varnished slip of wood, measuring 8 cm. by 4½ cm., covering it afterwards by means of a watch-glass in such a way that the edges of the latter firmly adhere to the slip.

Larger dry preparations I mount for my collection in two ways. If the object has been so prepared that the parts intended to be shown can only be seen on one side, such as the view of the cavum tymp. from above or below, and horizontal sections of the entire organ, etc., holes are drilled through the preparation at its two opposite extremities, or at any other suitable point; it is then fastened by means of metal tacks to a small, black slip of wood, which may be polished or varnished. It is hardly necessary to say that the holes drilled in the object for fixing it should not be made in any part which is anatomically of importance, but, if possible, in places where the tacks are out of sight. To avoid their being noticed at all, a tack 1½-2 cm. long is driven half-way into the wooden slip, and the free portion of it having been brushed over with glue is inserted into a bore-hole drilled on the lower surface of the preparation. These slips vary in size, according to the bulk of the object, from 9 cm. by 6 cm. to 14 cm. by 8 cm.; for small objects slips of 6 cm. by 4 cm. are sufficient. Instead of small wooden slips, square or oblong glass plates may be used, on which the preparations are fixed by means of glue or glass cement (syndektikon). To protect the objects from dust and insects, from four to eight specimens are kept in correspondingly large cases, the lids of which may be partly of glass so as to render inspection more convenient. It is absolutely necessary that these cases should be made as air-tight as possible, especially for dry preparations, the soft parts of which have been impregnated with Laskowski's preservative fluid (glycerine 100'0, acid. carbolic. 5'0, acid. boric. 1'0), because such objects, if not sufficiently protected, become in the course of time so soiled, that it is scarcely possible to again perfectly clean them.

Preparations that require to be examined on all sides, such as the membrana tympani, with the ossicular chain and the labyrinth, prepared out in one piece, are mounted, as shown in Fig. V., p. 191, on a stand which may be covered with a bell glass of corresponding size. The diameter of the round stand measures 10 cm., the height of the upright piece supporting the preparation 4 cm., and that of the shade fitting into a groove in the stand 8-9 cm.* This mode of mounting, on account of its rendering it possible to inspect the object on all sides, is also to be recommended for those corrosion preparations which have been treated by the wax and resin process, and require to be covered with a glass shade to protect them from dust and mechanical influences.

Small dry preparations, tympanic membranes, ossicula, casts of the labyrinth, may also be mounted on black slips of wood under glass and frame. A model collection of preparations of the ear is that which represents the comparative anatomy of the ossicles and osseous labyrinth, and which has been mounted on two tablets by Hyrtl's master-hand.

PROTECTION OF DRY PREPARATIONS AGAINST DESTRUCTION BY INSECTS.

All dry preparations intended for preservation must, in order to protect them against destruction by the larvæ of insects, be brushed over with a solution of glycerine sublimate (3 parts glycerine, 7 parts water, 0'01 sublimate of mercury). Instead of the sublimate, arsenious acid (0'1) may be employed. The painting over of the soft parts (not the bones) is done with a fine camel's-hair brush. Especially the outer and inner surfaces of the membrana tympani, which is most frequently destroyed by the larvæ, should be carefully pencilled over, so as to become impregnated with the fluid. In order to preserve the pliancy of the membrane, and to prevent it from cracking, it is advisable to repeat the pencilling over after the lapse of 3-4 years. Tympanic membranes on preparations which have been saturated with Laskowski's preservative fluid (5% carbolized glycerine, with the addition of 1% boracic acid), and which, as a rule, are not attacked by the larvæ of insects, should, for greater safety, be treated in the same manner.

* In this way I have mounted a collection consisting of sixty preparations, which were purchased in 1876 for the Museum of the College of Physicians in Philadelphia.

I.

II.

III.

IV.

V.

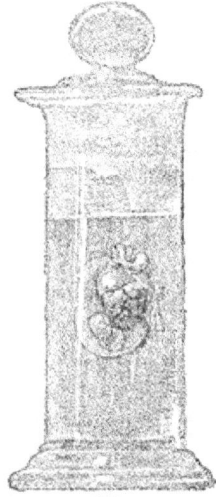

VI.

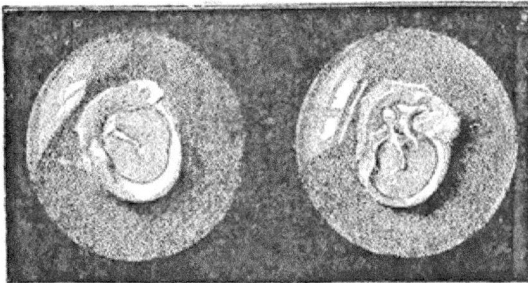

VII.

In spite of these precautions for protecting the preparations from destruction, I consider it advisable at the beginning of the warm season to provide the various cases of the anatomical collection with pieces of camphor wrapped in gauze, thus preventing their being infested with insects.

Preparations showing the presence of insects by such traces as brownish-gray granules, must either be entirely separated from the collection, or, for the purpose of destroying the larvæ, be put in sublimate spirit (1:300). If several specimens of the collection show signs of being affected, all the dry preparations should be placed in an air-tight box in order to destroy the larvæ by exposing them for five or six hours to the fumes of carbon bi-sulphide.

SECOND PART.

THE HISTOLOGICAL EXAMINATION OF THE ORGAN OF HEARING IN ITS NORMAL AND PATHOLOGICAL CONDITION.

INTRODUCTION.

THE histological examination of the tissues of the auditory apparatus forms an essential part in the dissection of the ear. For although, by dissecting pathological organs as macroscopic objects, we frequently enough gain information as to the anatomical cause of the disturbance of hearing, yet in many cases it is only the histological examination which will throw sufficient light on those functional impairments which originate from the finer structural alterations in the auditory apparatus, and which it would be impossible to recognise with the naked eye.

The histology of the organ of hearing is beset with far greater difficulties than that of other organs. This is due to the fact that it is made up of tissues with greatly varying peculiarities. Thus we see the sound-conducting apparatus and the delicate terminal structures of the acoustic nerve in the labyrinth surrounded on all sides by exceedingly dense osseous envelopes, which admit of a pathologico-histological or a topographico-anatomical investigation only when, by proper methods of decalcification and embedding, all the tissues are rendered fit for sectioning. These preparatory methods, however, require the greatest care and circumspection, in order to prevent the delicate structures of the connective tissue, of the epithelium, and of the nerves in the middle and inner ear from being altered and injured by the action of the decalcifying fluids.

In nearly all cases, therefore, must organs of hearing, intended for microscopical examination in their normal as well as their pathological state, be treated with certain chemical agents in order to preserve, as far as possible, the natural structure of their cells and tissues (fixing); to prevent the setting in of decomposition, which might have a destructive influence on their delicate texture (preservation); and to give their dissimilar tissues a homogeneous consistence, so as to render them fit for sectioning (hardening, decalcification). Where it is merely

13

a question of examining the soft structures of the ear (tympanic membrane, lining membrane of the meatus and middle ear, Eustachian tube, membranous labyrinth), it is generally possible to comply with all the above-mentioned conditions by making use of a suitable preservative fluid (chromates, alcohol), which effects both hardening and fixing, and prevents putrefaction.   In some cases, however, where special forms of organization (cell-nuclei, nerve-fibres, etc.) are to be examined, it is necessary to have recourse to a combination of fixing and preserving methods. It is, of course, understood that, in cases where a topographico-histological examination of the organ is intended, the parts selected, or even the entire apparatus, with its osseous envelopes, must be decalcified before being submitted *in toto* to microscopical examination.

PREPARATORY METHODS.

1. FIXING AND HARDENING.

The object of fixing is the perfect preservation of the natural cell-structure. The normal histology of the ear, especially the complicated anatomical relations of the membranous labyrinth, cannot be successfully studied without previously fixing the object.   Particularly where the chief aim is to give prominence to special protoplasmic cell-formations in the microscopical image the finest methods of histological treatment must be employed, in order to demonstrate not only the microscopic structure of the exceedingly delicate portions of the auditory apparatus, but also to bring under observation those of the more important protoplasmic formations therein which may be chosen for the purpose.   Besides the usual methods of fixing the preparations, the impregnation of the objects with metallic salts (osmium, gold) is largely employed, whereby—especially the nerve-elements of the internal ear—the cellular terminal expansions of the acoustic nerve, by virtue of their peculiar chemical reaction upon these substances, present themselves in a distinct form and colour.

The fresher the object, the more natural are the representations obtained by the fixing.   Successful metallic impregnations, in particular, can only be obtained on perfectly fresh organs.   The gilding methods used in neuro-histology (Freud) can only be employed to a limited extent in oto-histology, because the decalcification of the object, which must necessarily precede, commonly deprives the cell-protoplasm of its reducing power.

It must be accepted as a general rule that the organs of hearing under treatment, after removal of any superfluous portions still attached to them, are to be immersed in the liquid direct, without having previously been washed. Most of the substances here employed being equally suitable as hardening, preservative, and even decalcifying media, the organ under preparation should be kept in the solution until sufficient hardening or decalcification of the soft parts has been effected.

1. *Chromic Acid and its Salts.* — Chromic acid in 0.1-1% aqueous solution has fixing, hardening, preserving and decalcifying properties, and forms therefore an excellent medium for our purposes. The only disadvantage it has, though only when kept in action too long, is that the preparations frequently assume an intense dark-green colour, through the reduction of the oxide of chromium, in consequence of which the soft parts of the organ are rendered unfit for staining with carmine, hæmatoxylin, and aniline dyes.

Most in use as fixing media are potassium and ammonium bichromicum in a 2% solution, and Müller's fluid, a mixture of 2 parts bichromate of potash, 1 part sulphate of soda, and 100 parts water. If these solutions be used in sufficient quantity and changed every day, we succeed in durably fixing the tissue structures of the ear in from one to three weeks, sometimes even in a few days, after which the decalcification can be proceeded with.

2. An exceptional reputation in oto-histology has Tafani's fixing fluid. It consists of 80 parts of a 0.4% aqueous solution of bichromate of potash and 20 parts of a 1% aqueous solution of hyperosmic acid. The two solutions are only to be mixed immediately before use. Leaving the preparations in the solution for twenty-four hours is sufficient to fix them, and to reduce the osmic acid.

3. Good results are also obtained by employing Vlakovic's less costly chrom-alcohol fixing fluid, which I frequently use for fixing pathological preparations; it consists of one part of a 5% aqueous solution of chromic acid, and 16 parts of 90% alcohol, and effects sufficient fixation in twenty-four hours. After-hardening in alcohol.

A fixing fluid which is greatly to be recommended for normal and pathological organs is that of Urban Pritchard: acid. chromicum cryst. 1 g., aqu. dest. 20 g., alcohol 90° 180 g., in which the object is left for seven to eight days.

4. For pathologico-histological investigations, especially where it is intended to study pathological cell-proliferations, Flemming's fixing method, now generally employed to show nucleus segmentation, is specially to be recommended. This solution consists of 15 parts of 1% chromic acid, 4 parts of 2% osmic acid, and 1 part of glacial acetic acid. The tissues, which should be as fresh as possible, are left for two or three days in the solution, and having been well washed,

they are decalcified, then hardened in alcohol, sectioned, and coloured with nucleus-staining aniline dyes (see Staining Methods).

5. Good results are also obtained from the 3% nitric acid medium, recommended by Altmann, which, on account of its excellent decalcifying properties, is well suited for oto-histological purposes. (The subsequent staining is to be effected with hæmatoxylin.)

6. *Impregnation with Metallic Salts.*—For oto-histological purposes the treatment with gold and osmium is to be preferred. By the action of the gold and osmium compounds on the tissue, not only an excellent fixing of the structural elements is obtained, but, what is of special value for the histological examination of the nervous and sound-conducting apparatus, the metal, owing to the reducing properties of certain tissue elements (nerves, epithelial cells, etc.), is precipitated in a fine granular form, while these elements themselves appear sharply differentiated in the microscopical image. Metallic impregnations are, therefore, chiefly employed for the representation of the vessels and nerves of the membrana tympani, of the lining membrane of the tympanic cavity, and of the Eustachian tube, as well as of the nerve terminations in the saccules, ampullæ, and in Corti's organ. Those portions of tissue which are to undergo metallic impregnation should, in as fresh a state as possible, and without previous treatment in any other way (alcohol, chromium, etc.), be immersed in the fixing fluid, which must be kept in a brown bottle, and not exposed to the influence of the light.

I. *Gilding.*—The re-agent exclusively used is chloride of gold in a ½-1% aqueous solution. The action of the metallic salt solution having taken effect, a weak solution of an organic acid (formic or acetic acid) is employed to accelerate reduction.

The peculiar effect of the chloride of gold is that the axis-cylinders of the nerves and nerve terminations in the labyrinth assume a colour from dark violet to deep dark-red. The vessels and nerves of the tympanic membrane, and of the lining membrane of the middle ear, the cell protoplasm of the various epi- and endothelia (mucous membrane, tympanic membrane, membranous labyrinth, etc.), show the same colour, but only when examined in a perfectly fresh state.

Methods for gilding : (*a*) The fresh organs, or portions of them, are first of all placed in a 0.5% solution of chloride of gold. The length of time they are to remain therein depends on the kind of tissue · under treatment, and on the bulk of the preparation. Membranes, drumhead, lining membrane of the tympanum and of the Eustachian tube, saccules, and semicircular canals are left in it for a quarter of an hour, while more bulky tissues, tuba Eustach., tympanum, and the pyramid, require from half an hour to an hour. Then the gold preparations are exposed to the light, and for twenty-four hours kept in a fluid which contains 10 drops of formic acid for every 10 g. of water. With tissues containing much calcareous matter, the gold solution must be taken stronger; for Corti's organ, of about 5%. The objects assume on the surface a deep dark-violet colour.

(b) Another method is to place the parts of fresh organs in a 0.5% solution of chloride of gold, to which is added a minute quantity of acetic acid, and to leave them in it for a quarter of an hour to an hour, until they acquire a straw colour. They are then washed in distilled water acidulated with acetic acid, and kept for twenty-four to forty-eight hours standing in a light place, until reduction has fully set in. The colour of the preparations has then become deep bluish-red, violet, blue, or dark gray. After-hardening in alcohol.

(c) Dr. E. Berger, of Paris, gilds after the following method, which I have employed with advantage on Corti's organs, taken from young animals immediately after they had been killed. The objects are immersed in a solution the composition of which is as follows: 1% chloride of gold, 1% hydrochloric acid, 23% glycerine, and 75% distilled water. In this the preparations remain six to eight hours, until they become of a dark-blue colour; then, washing in water, and after-hardening in alcohol. For decalcification I used a saturated solution of picric acid.

(d) Good reduction is also obtained by the use of amylic alcohol (1 drop to 10 g. water), which, after impregnation of the object with gold, is used either alone or combined with acetic or formic acid as a reducing agent.

Stöhr boils 8 ccm. of a 1% solution of chloride of gold with 2 ccm. of formic acid in a test-tube, and then allows it to cool. In this mixture the object remains for an hour in the dark; it is then washed in distilled water and immersed in a mixture of 10 ccm. formic acid and 40 ccm. distilled water, and exposed to the air. Reduction is effected in twenty-four to forty-eight hours. The preparation to be kept for eight days in the dark in 70%, then 90%, alcohol.

Impregnations with gold or with osmium prove effectual only in the superficial layers, since the chemical changes resulting from the peculiar reaction in the tissues prevent the metallic salt solutions from penetrating deeper. To obtain successful representations, care should above all be taken to bring the tissues which are to be gilded into sufficient contact with the chloride of gold solution. When examining the lining membrane of the tympanum, and the membranous structures of the labyrinth, these cavities must be opened in a suitable manner without doing injury to important parts. For the tympanic cavity this is done at the tegmen tymp.; for the labyrinth, at the superior semicircular canal, or else by piercing the membrane of the fenestra rotunda.

If it be merely intended to gild the tympanic membrane or the lining membrane of the tympanum and of the Eustachian tube separately, it is necessary, in order to insure direct action of the metallic solution, to remove any accumulated secretion and epithelial deposit from their exposed surfaces, by brushing them over with a solution of common salt (0.6%).

The after-hardening is effected in alcohol, decalcification in the case of petrous bones of slight consistence (young animals, embryos) by picric acid; with consistent ones, by an alcoholic solution of hydrochloric acid or nitric acid (see Methods of Decalcification).

II. *Osmium Treatment.* Osmic acid in a 0.5-2% solution for fixing and reducing fresh tissues of the organ of hearing is made use of, where it is intended to represent bloodvessels, medullated nerve fibres, and pathological fatty degeneration in cellular elements. As the red blood corpuscles become dark brown under the action of osmic acid, the vessels filled with them stand out as strongly marked, deep, blackish-brown strands from the rest of the tissues. All fat-containing tissues assume under osmium reduction a blue-black colour, so that the nerve medulla and fat-cells can be made easily distinguishable. Like the treatment with chloride of gold, osmium reduction is limited chiefly to the surface of the preparations, which makes it necessary to use all the precautions observed in gilding. The staining caused by the osmium reaction shows better on fresh preparations than on those previously hardened, especially when treated with alcohol, because the latter deprives the tissues of part of their fat.

Directions.—Structures of this kind which are to be examined in part or as a whole must, according to their bulk, be placed in a small quantity (1-8 ccm.) of a 1-2% aqueous solution of osmic acid, and left in a dark place, until they are hardened and have assumed a blue colour (1-24 hours). After this, repeated thorough washing for several hours in distilled water, which is to be frequently changed, and to which a few drops of acetic acid should be added; then washing, for a like space of time, in a 0.6% solution of common salt. Soft parts are to be preserved in alcohol, which should be gradually increased in strength; bone must be decalcified according to one of the methods to be described further on.

III. *Chloride of palladium* acts on the nerves and protoplasms in a similar manner to that of chloride of gold. A combined osmium-palladium method for fixing, and colour reduction of the membranous labyrinth, we owe to Dr. Schönlein, of Halle, the directions for which are here given : The cochlea having been prepared out, after removal of the stapes from the fenestra ovalis, is placed in a $\frac{1}{2}\%$ solution of osmic acid, and left in it for twenty-four hours. The preparation is then washed and immersed in a 0.1% solution of chloride of palladium, to which ten drops of hydrochloric acid are to be added. This solution serves also for decalcification. It must be well shaken several times a day, and one drop of hydrochloric acid added each day until the decalcification is complete. The process requires from three to six weeks, according to the hardness of the bone. Then follows thorough washing, after-hardening in absolute alcohol, embedding in celloidin, and sectioning.

The chromates, described in detail at the commencement of this chapter, effect also the hardening of the soft parts of the organ of hearing to that degree of consistence which is necessary to render them fit for cutting. The greater the quantity of fluid used, and the oftener it is changed, the more perfect will be the hardening. Owing to the drawback of frequently occurring coagulation images, the rapid method of hardening with absolute alcohol, which, if changed two or three times in the course of the day, will render the preparations perfectly hard and fit for cutting, is rarely used in oto-histology.

When sufficiently hard, the object is thoroughly washed in distilled water ; for perfect hardening and further preservation it is placed either in alcohol gradually increased in strength, or, for the purpose of decalcifying it, in a suitable fluid (see Decalcification).

2. DECALCIFICATION.

To decalcify the osseous structures of the ear, aqueous solutions of different mineral acids in various degrees of concentration are exclusively used.

There are two methods of decalcification, one which occupies more time and is more preservative, and another (the so-called rapid method) which requires the acids to be used in a higher degree of concentration. Each of these two methods has its advantages and disadvantages, which shall be duly considered in connection with the description of the various methods of decalcification.

Every object which is to be decalcified, after having previously been subjected to a suitable process of fixation, is, before its immersion in the decalcifying fluid, washed in fresh water until the latter ceases to be coloured. Next, the osseous portion of the object is, as far as possible, freed from adherent soft parts, the dura mater in particular to be stripped off from the petrous bone, so as to allow the decalcifying fluid to act directly on the bone. The preparation must, by removal of all superfluous osseous parts with saw and bone forceps, be reduced to the smallest possible bulk, and placed in a sufficient quantity of fluid, since decalcification is the more uniform and rapid the smaller the object, and the greater the quantity of the medium used in the process.

The proceeding must be continued until the preparation has become soft throughout, and thus fit for cutting. The best test for this is to thrust a fine preparing needle at various points into the petrous bone. If this can readily be done in many places, and if, especially in the osseous mass behind the porus acust. internus, no hardness be found (Moos), it may be assumed that the petrous bone is in such a condition as to be ready for cutting. If, however, on the needle being pushed in, resistance should be encountered in the deeper parts, this would show that there are still undecalcified portions, and the preparation must be left to the action of the fluid until perfectly decalcified.

Methods of Decalcification.

1. Müller's fluid, which has been repeatedly mentioned, and the excellent qualities of which, for fixing and hardening tissue, have already been pointed out, may, on account of its containing free chromic acid, be advantageously used as a decalcifying fluid, but only for small osseous structures (ossicula), or comparatively soft temporal bones (of embryos, new-born infants, young animals, etc.). For the decalcification of petrous bones of adults it is not suitable, as the quantity of free chromic acid which it contains is only trifling. Even with small bones, or those containing little calcareous matter, it is necessary to use the liquid in abundant quantities, and to frequently change it (at first every day, later every third or fourth day). Five to ten drops of concentrated nitric acid, added at the end of the first week to half a litre of Müller's fluid every time it is changed, will, without causing any injury, considerably shorten the process, which, even with petrous bones of slight consistence, such as those of very young individuals, takes weeks to complete.

For the histological investigation of organs of hearing taken from embryos, children, or young animals, this method has its advantages, because it preserves the elements of the connective tissue and nerve-structures in a perfect condition, and the nerve-elements of the sound-perceiving apparatus in the labyrinth admit, when decalcification has been completed, of staining after one of the usual nerve-staining methods (Weigert's hæmatoxylin.).

2. *Chromic Acid Method (Waldeyer, Gottstein, Moos).*—For decalcifying the petrous bone Waldeyer recommends an aqueous solution of chromic acid, with the following directions : The superfluous soft parts and osseous portions having been removed, the object is first placed in a weak solution of chromic acid (1 : 600), remains there for several days, is then transferred to another solution of the same acid of 1 : 400 water, and after a similar interval to a stronger solution of 1 : 200 water. This solution is now used as the decalcifying fluid proper, and is to be renewed every five or six days, when, to shorten the process, 2 g. of officinal concentrated nitric acid are added to every 100 g. of the fluid. After perfect decalcification, which with petrous bones of adults takes three months to complete, the preparation is quickly washed in water and placed for a few hours in spirit. Only then is it thoroughly washed and freed from acid by placing it for some hours under a stream of water until the latter flows off quite colourless. The further proceeding is to immerse the object in spirits of wine, the strength of which is to be gradually increased, absolute alcohol being used a few hours before sectioning.

As a rule good results are obtainable by this method, as it has an additional advantage, the chromium being an excellent hardening agent for the soft parts. It has, however, one drawback, viz.: that in the vicinity of the internal meatus there frequently remain undecalcified portions of bone (Moos) which, when the rest of the preparation is already soft, cannot be got rid of without injury to the entire object, an exact histological examination thus becoming impossible. That the chromic and nitric acid decalcification frequently causes alterations in the labyrinth, in Corti's and Deiters's cells, in the tunnel fibres, and Reissner's membrane (Katz), which may easily be mistaken for pathological changes, will be confirmed by those who have been making use of this method for some time. Besides, through want of sufficient care in the proceeding, it may happen that the objects become too soft and crumble away into a useless mass.

3. *Picric acid* in a saturated aqueous solution is well suited for the decalcification of the petrous bones and ossicula of embryos and children. When the study of the normal histological structures of the labyrinth is intended, the decalcification by picric acid may be preceded by the gilding of these parts. To decalcify the hard petrous bones of adults picric acid is not to be recommended, on account of its comparatively slow action.

4. *Nitric Acid Decalcification.*—The fluid used in the process is made up of 100 parts water, $3\frac{1}{2}$ parts nitric acid, and $\frac{3}{4}$ of a

part chloride of sodium. The solution, which is to be used in large quantities, is at first changed every day. Normal organs, on which the mucous membrane of the middle ear or labyrinth are to be examined, must before decalcification be fixed in Müller's fluid or in a solution of osmic acid. In pathological cases I allow, as a rule, fixation in Vlakovic's solution to precede decalcification by nitric acid (see p. 195). But for the examination of the external and middle ear it is often sufficient to place the object in the medium direct, as a 3% solution of nitric acid alone is an excellent means of hardening, and often used for histological purposes. Decalcification and softening are effected by this method in a perfectly safe and uniform manner without detriment to the soft parts of the organ. For the study of the pathological condition of the ear the nitric acid method is of special value, as by the action of the acid the karyokinetic figures are rendered visible. The decalcification of a human temporal bone is completed by this method in twenty to twenty-five days. The preparations are then well washed and preserved in 50% alcohol. By rectified or absolute alcohol the decalcified bone is hardened to such a degree that it must be placed in warm water, or in warm dilute alcohol, to restore its consistence sufficiently to render it fit for cutting.

5. *Hydrochloric Acid Decalcification.*—For decalcifying the organ of hearing, hydrochloric acid should not be used in an aqueous, but in an alcoholic solution, on account of its causing the soft parts to swell out. Besides, it is advisable to add a small quantity of chloride of sodium to the acid solution. For the histological examination of the labyrinth the hydrochloric acid method is not to be recommended, because of its injurious action upon the labyrinthine structures, the macul. acusticæ, and Corti's organ, but as a rapid method when making topographical sections of the entire organ. We employ for our purpose a mixture consisting of 5 to 15 parts of concentrated hydrochloric acid, 1,000 parts rectified spirit, 200 parts water, and 5 parts common salt. The latter counteracts the swelling out of the connective tissue caused by the hydrochloric acid. Large quantities of the liquid, and its renewal from day to day, are absolutely necessary if rapid and good results are to be obtained The washing and preserving of the decalcified organs to be carried out as in the previous method.

3. EMBEDDING AND SECTIONING.

When, by the proceeding explained above, the tissue-elements have been fixed and hardened, and the entire organ has been reduced, by the decalcification of its osseous parts, to a consistence which admits of cutting, the preparation must be embedded in the regular way, so that, during sectioning, the delicate structures of the middle ear and of the labyrinth may be retained *in situ*. Embedding is indispensable, not only when making sections of the tympanic membrane, of the articulations of the ossicula, and of the labyrinth, but also where it is intended to cut the entire organ into serial sections for topographico-histological and pathologico-anatomical purposes. It is obvious that the sound-conducting and sound-perceiving organs, suspended in the cavities of the temporal bone (tympanic membrane, ossicula and their articular connections, membranous labyrinth, saccules, ampullæ, membranous semicircular canals, Corti's organ, can only be perfectly preserved *in situ* on sections, if these cavities have been previously filled with some medium which, at first fluid, but stiffening afterwards, will retain all the parts in their natural positions. If this is not done, some of the structures only loosely connected with the osseous walls of the middle ear or of the labyrinth might fall out of the section. Besides, these delicate structures are so liable to injury that, in order to prevent their being torn, it is requisite to entirely fill up the cavities with a medium which, though sufficiently firm, will admit of cutting, and allow, even in sections, all parts to remain in their normal connection and position.

The older embedding masses—among which we mention mucilage (hardening in alcohol), glycerine glue (fluid when hot, firm at room temperature), paraffin (5 parts paraffin, 2 parts spermaceti, 1 part pure hog's lard), the mixture of oil and wax, the white of an egg—are, with the exception of paraffin, which is still made use of in some cases, of little value in oto-histology, since we have been made acquainted by Schiefferdecker and Duval with colloidin, an embedding medium which answers the purpose in every respect.

Colloidin, a substance resembling collodion, is procurable in the trade as white transparent cakes, and has the property, when mixed with ether and absolute alcohol (in equal parts), of gradually dissolving into a syrupy fluid, which, however, again solidifies as soon as the dissolving medium is allowed to evaporate, or if dilute alcohol be added.

1. Formula for the celloidin solution.

For oto-histological purposes, three solutions in different degrees of concentration should always be kept in readiness: one thin-fluid, another somewhat stiffer, and a thick-fluid one. The first is prepared by placing 30 g. of chemically pure celloidin, cut into small cubes, in a vessel hermetically closed by a glass stopper, and containing a mixture of 30 g. absolute alcohol and 30 g. sulphuric ether.

The second solution is obtained from a mixture of equal parts

(25 g.) of alcohol and ether, in which 30 g. of celloidin are dissolved. Finally, the most concentrated solution is made by dissolving 30 g. of celloidin in alcohol and ether (20 g. of each). With frequent shaking of the mixture, the solutions of the required consistence (oil or thick syrup) will be ready in a few days, and are to be used according to the directions which shall be given later on.

2. *Preparation of the structures of the ear for celloidin embedding.*

(a) The completely decalcified objects are first thoroughly washed in distilled water, which must be frequently changed, until they are perfectly freed from the acid.

After this the tympanic cavity, semicircular canals, and cochlea should be opened according to the methods described, so as to allow the embedding substance to penetrate into the cavities of the temporal bone.

(b) Then the preparation must be thoroughly dehydrated, because the celloidin could not enter any tissues, or tissue spaces, containing water. For this purpose the object is placed for twenty-four hours in dilute, and afterwards from one to two hours in concentrated, alcohol. When structures of the external and middle ear are under examination, the specimen may be transferred direct from the absolute alcohol to the celloidin solution. In the case of the labyrinth, however, where it is absolutely necessary that all the spaces should be completely filled with the medium, the preparation is taken from the alcohol into a mixture of 6 parts sulphuric ether and 1 part absolute alcohol, where it is to remain for at least twelve hours.

(c) After this the preparation is placed in a weak solution of celloidin, where it remains from one to three days, according to the size of the object. Preparations of the external and middle ear may be transferred from this solution direct to the thick-fluid celloidin mass. Those of the labyrinth, however, in order that the embedding may be the more successful, are taken from the weak solution into a more concentrated one, and from this into the thick fluid.

3. The organs, permeated with the celloidin solution, are taken out at the end of twenty-four to thirty-six hours, and placed in a small paper case of corresponding size, which has been previously filled to about one-third with partially-set celloidin. Then the most concentrated solution of the medium is quickly poured over the preparation, and left for two to three hours under a bell-glass to allow it to thicken by evaporation. After that the embedded object is transferred to 80%, and at the end of five to six hours to 70%, alcohol, where, in the course of twenty-four to forty-eight hours, the whole mass will set, and become fit for cutting.

Before embedding the preparations it must be determined whether they have not been over-hardened by the absolute alcohol, in which case they must be softened again in warm dilute alcohol. This proceeding should also be observed when the object has become over-hardened by the celloidin embedding itself, which frequently happens with preparations decalcified in the nitric acid and salt solution.

In order not to miss the direction of the cut when making sections on embedded preparations, that surface which in the microtome comes to lie upwards is, before embedding, brushed over with carmine, which shines through the transparent medium.

If the embedding is perfect, the celloidin, when set, must in the microscopic section completely surround all the tissues, and retain them *in situ.* The medium being quite transparent in thin sections, it in no way interferes with the microscopical examination. The celloidin may easily be removed from the sections by placing them in a mixture of alcohol and ether for a quarter of an hour, clearing up being effected with oil of cloves. This should, however, not be done under any circumstances, when it is intended to examine those structures of the middle ear (ossicula, bridges of connective tissue) and in the labyrinth (saccules, ampullæ, semicircular canals, and cochlear membranes) which are only loosely connected with the osseous walls, because by dissolving the fixing medium these structures would fall out of the section. On the other hand, where the various parts of the organ are to remain *in situ,* care ought to be taken in any proceeding with the sections that the fixing substance be preserved. When sectioning the preparations, the blade of the knife must, therefore, not be moistened with absolute alcohol, but common spirit should be employed.

Photoxylin, which in recent years has been recommended as a substitute for celloidin for embedding purposes, has, in my own oto-histological investigations, not proved so useful. It certainly possesses the advantage of dissolving in a few minutes in alcohol and ether, but it has the great drawback that, during dehydration of microscopical sections in absolute alcohol, the photoxylin layer easily dissolves, and the preparation falls in pieces.

Mention might here be made of the rather complicated method of paraffin embedding, as practised in the laboratory of Professor Klebs, of which I myself have no experience, but which, according to the statements of Dr. Rohrer, of Zürich, offers many advantages for special examinations of the membrana tympani and of the labyrinth.

(1) The previously fixed and decalcified preparation is, for the purpose of dehydration, placed in absolute alcohol.

(2) From the alcohol the preparation is transferred to pure aniline oil, in which it remains at a constant temperature of 40° to 60° for twenty-four to thirty-six hours, until the object becomes of a glassy transparency.

(3) After this the specimen is placed for twenty-four to thirty-six hours in xylol, which during that time must be twice renewed, until the aniline oil has been completely got rid of.

(4) Next the preparation is transferred to xylol, slightly mixed with paraffin; and finally

(5) Into pure melted paraffin of 56° C., under which temperature it is to remain for forty-eight hours.

• When the paraffin mass has become cold and set, the preparation is taken out and cut in the microtome.

The sectioning is carried out in the usual way. On the whole, cutting with a well-constructed sliding microtome is for our purpose much to be preferred to all other methods, especially on account of the evenness obtainable in the sections and there being no loss of material, while the whole manipulation is more expeditious. Where serial sections are required, this method is altogether indispensable. For several years I have been using the automatic sliding microtome of Reichert of Vienna; on rare occasions one of his hand-microtomes, which for small objects and for less minute examinations are quite sufficient.

The decalcified osseous tissue offering greater resistance to the knife of the microtome than hardened soft parts, most preparations must, during the sectioning, be fixed in clamps, because fastening them simply upon cork with celloidin, as was formerly done, does not give the object sufficient firmness.

4. STAINING OF THE SECTIONS.

We employ for oto-histological purposes principally the following dyes and staining methods:

1. *Carmine Staining.*—Carminate of ammonia and Beale's carmine are specially adapted for staining decalcified bones and cell-proto-plasms, without colouring the nuclei. By leaving the specimens for twelve to twenty-four hours in a greatly diluted rose-coloured carmine solution (slow carmine staining), the sections take a beautiful colour (Stöhr).

*Picro-carmine** is chiefly employed for pathologico-histological examinations of the organ of hearing (pathological ossifications, de-

---

* Carmini, Ammon. pur. liq. . āā 0.50, Aq. dest. 25.0 ; cui digerendo adde Acidi picrin. conc. puri 75.0 et ūltra.   Dein adde Phenoli puri gtts. duas.

structive and tuberculous osseous processes in the temporal bone, round-cell proliferation). It stains the osseous tissue brick-red ; the medulla and cell-protoplasm, as well as the muscular fibres, take a diffuse yellow colour; while the cell-nuclei appear cherry-red.

Staining for five minutes, then immersion for half an hour in a hydrochloric solution of glycerine (1 : 100), then washing in water and alcohol.

For rapid and durable nucleus-staining, alum-carmine* is also to be recommended (Grenacher), as it offers the additional advantage that the sections, even when left for a considerable time in it, do not become over-stained. Five to ten minutes are sufficient to give a perfectly distinct colour.

2. *Eosin* in a 5% solution, poured drop by drop into a watch-glass filled with distilled water or alcohol in which the sections have been placed, gives in a few minutes a bright-red colour, without differentiating nuclei and cells (rapid staining). We therefore employ it as a ground-colour, to be followed by a suitable differentiating dye for the nuclei. This is effected, after thorough washing of the sections in alcohol, by after-staining in hæmatoxylin, methylene blue, or gentian violet.

3. *Hæmatoxylin.*—We prefer the hæmatoxylin-alum stain (hæmatoxylin 2.0, alcohol absol. 100.0, aq. destill. 100.0, glycerin. pur. 2.0, alum. pulv. 2.0), as mentioned in Friedländer's microscopic technique. To be filtered every time before it is used. Does not stain well until a week after it has been prepared. In a few minutes the sections are coloured brown, but change to blue in distilled water, in which they must be washed for a quarter of an hour to two hours. The dye gives the nuclei, the decalcified bone, and its vessels a beautiful blue.

Bacteria-Staining on Sections.

To show the presence of micro-organisms on sections of the auditory organ, after it has been preserved in absolute alcohol and decalcified by nitric or hydrochloric acid (see Decalcification), we recommend Löffler's 'universal bacteria-staining' method. When perfectly free from acid, the sections are left for five to ten minutes in an alkaline solution of methylene blue (30 ccm. concent. alcoholic methylene blue solution to 100 ccm. 0.01% solution of caustic potash). Wash for a few seconds in 1% acetic acid, then place in alcohol, clear up with cedar-oil, and mount in Canada balsam. All schizomycetes, especially diphtheric micro-organisms and pyogenic cocci, as well as the mycelia of mould-fungi (aspergillus), take a beautiful blue.

* Carmini 1.0.   Alumin. pulv. 5.0.   Aq. dest. 95.0 leni calore solvet et filtret.

Bacteria may also be stained separately by Gram's method. Staining in aniline-gentian ; decolorization by alcoholic solution of iodine.

For the demonstration of tubercle-bacilli in sections of carious portions of the temporal bone, we particularly recommend the staining method suggested by Ziehl-Neelsen. The solution employed is the so-called carbol-fuchsin,* in which the sections remain fifteen to twenty minutes. Decolorization with 25% sulphuric acid, washing in rectified alcohol ; after-staining (three minutes) in a watery solution of methylene blue. Bacilli red, cells and tissues blue.

5. CLEARING AND MOUNTING OF THE SECTIONS.

The further treatment of the sections after staining, up to their mounting under glass, is carried out after the following well-known methods : (1) Dehydration of the sections in absolute alcohol, according to their size and thickness, for five to fifteen minutes. (2) Careful lifting out of the sections from the alcohol by means of a metal spatula bent at a right angle to the handle, and drying with fine blotting-paper. (3) Clearing of the section.   With preparations embedded in celloidin, this must be done either by steeping in origanum oil of the finest quality, or in aniline oil, ol. thymi., cedar oil.   On the other hand, the use of oil of turpentine and oil of cloves must be avoided, as these, through dissolving the celloidin, cause a milky cloudiness of the object, and dislocation of the more delicate structures. With paraffin embedding, however, the last-named oils are of great service, in so far as the preparations are not only cleared, but the paraffin is also dissolved, without the transparency of the sections being thereby diminished. (4) Transferring the preparation with the metal spatula on to the glass slip, where, after mopping up the ethereal oil with blotting-paper, the sections are mounted in Canada balsam. Thick, not sufficiently transparent sections are cleared by adding a drop of xylol to the balsam ; generally, however, xylol (only to be employed in paraffin embedding) renders the objects too transparent, which sometimes makes the differentiation of the tissues a matter of greater difficulty.

Sections which are to be preserved in glycerine are first of all washed in distilled water, then mounted in somewhat diluted glycerine, and the edges of the cover-glass cemented with black japan or white Frankfort lac. Celloidin sections do not alter in glycerine, but those embedded in paraffin cannot be mounted in that medium. Sections are now generally mounted in Canada balsam or Dammar, less frequently in glycerine. For the study of pathological changes in

* Rp. Fuchsin 1.0, Alcohol. absol. 10.0, Aq. carbol. 5% 100.0.

the structures, however, it is recommended to mount a few sections of a preparation in glycerine also, because some details stand out more clearly in that medium than in Canada balsam or Dammar.

6. SERIAL SECTIONS.

For the making of serial sections, particularly of the smaller structures—cochlea, semicircular canals, transverse sections of the membrana tympani, malleo-incudal articulation, system of cavities between malleus and incus and external tympanic wall, mucous lining of the tympanum, etc., where it is of importance to deduce from the series of sections the topography of the parts concerned—it is advisable to make use of the following method, practised at the zoological stations for the study of the smaller animals, and for the details of which I am indebted to Dr. Koller.

Staining *in toto* (1) in Ranvier's picro-carmine solution, (2) alcoholic hæmatoxylin solution, or (3) in borax-carmine.

The piece of tissue is transferred direct from the alcohol to the staining fluid, and left there, according to its size, for two to four days (in borax-carmine, one to two days). It is then washed (decolorized) in 70% alcohol (if borax-carmine be used, in acid alcohol). Duration of the decolorization, from a quarter of an hour to an hour.

Then the preparation is, for the purpose of dehydration, placed for several hours in absolute alcohol. It is next soaked in turpentine, after that removed to a mixture of turpentine and paraffin (on a water bath of 60°), is left there for several hours, and for a like space of time in pure (hard) paraffin.

After the embedding, the melted paraffin is made to set rapidly, by dipping in cold water, so as to give it a glassy, not white, snowy appearance.

The glass slip is painted over with as thin a layer as possible of equal parts of collodion and oil of cloves; the sections are placed upon it, and exposed for ten to fifteen minutes to a temperature of 60°, to allow the paraffin to melt and the oil of cloves to pass off. Thus, the section is firmly fixed upon the glass, which is then laid in a saucer containing turpentine, to extract the paraffin. The mounting in Canada balsam or Dammar is the same as with the other methods.

For making serial sections of some particular parts of the organ of hearing, a method* devised by Dr. Ignazio Dionisio, of Turin, is recommended, a description of which is here given. The microscopic sections are arranged in their order on the slip, and covered with a

* From the laboratory of Prof. Schenk in Vienna.

14

fine metallic net prepared for the purpose. By a special contrivance the net is fastened at both ends of the slip with two clamps, so that all the sections are fixed alike, none of them being pressed too hard. The whole of the preparations can be subjected to the usual treatment of microscopic sections, by placing the slip, still fixed in the apparatus, in the staining fluid, then alcohol, and, lastly, in oil of cloves or origanum oil. When the preparations have been sufficiently cleared in one of the last-mentioned oils, any superfluous oil is soaked up by gently pressing blotting-paper upon the net; this is then removed, and the sections are mounted in Canada balsam or Dammar.

This method has the advantage that even the thinnest sections do not curl up during treatment in the various liquids, and that the procedure of staining and clearing up, which with most of these methods takes a considerable time, is materially shortened.

Some special methods of the microscopic technique, in particular those employed in the examination of Corti's organ and the central course of the acoustic nerve, will be discussed in the following part of the work, having for its object the description of the histological examination of the various portions of the organ of hearing.

I.

THE HISTOLOGICAL EXAMINATION OF THE AURICLE AND OF THE CARTILAGINOUS MEATUS.

FOR hardening and preserving the auricle and cartilaginous meatus, Müller's fluid (p. 195) is best adapted. The preparation, having been left for two to three weeks to the action of the medium, is washed for an hour or two in distilled water, and hardened for three to four days in alcohol, which is to be gradually increased in strength. The parts to be examined are embedded either in celloidin or in paraffin, or sectioned between elder-pith, stained with carmine, eosin, picro-carmine or hæmatoxylin, after which the sections are mounted, some in Dammar, others in glycerine.

Sections of the normal auricle show: (1) The diameter (1-2 mm.) and the structure of the elastic cartilage, which proves to be fibro-cartilage. (2) The thickness of the perichondrium, the elastic fibres of which are connected, on the one side, with those of the auricular cartilage, and with those of the subcutaneous connective tissue on the other. (3) The structure of the integument, which shows all the elements of the cutis, with this difference, that on the convex side, particularly on the concha, the papillæ are very low, and that the numerous sebaceous follicles, with greatly dilated excretory ducts, empty themselves on the surface, while the sweat glands are here almost entirely wanting. Successful capillary injections of the cutis with carmine glue or Berlin blue yield beautiful representations of the arrangement of the bloodvessels, which may be traced from the cutis into the perichondrium.

The histological examination of the cartilaginous meatus is best conducted on sections made perpendicularly to its long axis. For this purpose the cartilagino-membranous portion of the meatus is carefully dissected away from the osseous, then either embedded in celloidin or filled with paraffin, and sectioned in the microtome in a vertical direction to its long axis. The ring-shaped sections bring

14—2

into view the relationship of the circumference of the cartilaginous
to the membranous portion of the meatus, and, with suitable staining,
yield a clear representation of the position of the hair-follicles, of the
acinose sebaceous glands, and of the ceruminous glands in the cutis
(Fig. 131). The sebaceous glands (o o') discharge for the most part
laterally into the hair-follicles (h h'), less frequently on the surface of
the cutis. Below the layer of the hair-follicles there lie, in the subcutaneous connective tissue, rolled up like a ball, the ceruminous

Fig. 131.—Section through the cutis of the cartilaginous meatus of a new-born infant.
h h' = hair-follicles ; m = opening of a hair-follicle on the surface of the cutis ; o o' =
sebaceous glands, discharging laterally into the hair-follicle ; c c' = section through the
coil of ceruminous glands in the deeper layers of the cutis ; g g' = excretory duct of the
ceruminous gland ; s = lateral discharge of this duct into the hair-follicle. After a
preparation of Dr. Gompertz.

glands (c c'), resembling the sweat glands. The diameter of the
glandular tube measures, according to Schwalbe, 0.1 mm., that of
the glandular body, 0.2-1.5 mm. Their excretory ducts (g g') are
narrow (0.01 mm.), and open, according to Stöhr, in children laterally

into the hair-follicles (s), in adults on the surface of the meatus, close to the hair-follicles. Their brownish-yellow colour is often retained, even on stained sections. If the vessels are successfully injected, the fine capillary network with which the ceruminous glands and hair-follicles are enclosed may be beautifully demonstrated.

In order to become acquainted with the arrangement of the hair-follicles and glandular elements of the cutis, and its epithelial lining, a series of cuts should be carried through the cutis, parallel to the surface of the meatus, consequently vertically to the long axis of the follicles and ceruminous glands. These have a cubical epithelium of a single layer (Stöhr), to which, externally, a layer of smooth muscular fibres and a membrana propria succeed. The epithelium is covered internally by a cuticular seam, which is especially distinct in adults.

Pathological preparations of the auricle and cartilaginous meatus intended for microscopical examination are subjected to the same treatment. But objects hardened merely in alcohol also yield useful sections. Suitable for simple alcohol hardening are especially : retrogressive changes in the auricle (age metamorphoses), and those peculiar connective-tissue proliferations and lacuna-formations as described by Pareidt and J. Pollak ; also the deposit of lime salts, gouty concretions, thickenings of the perichondrium, ossifications (Bochdalek, J. Pollak), etc. On the other hand, carcinomatous degeneration of the auricle, inflammatory infiltration of the cutis, abscesses in the cartilaginous meatus, ruptures and fistulous tracts in the incisuræ Santorini, and changes in the glandular elements of the cutis should always be fixed and hardened with chromic salt solutions, or Müller's fluid. It is to be understood that in those cases where sections are examined for the presence of micro-organisms, the method described on p. 207 must be followed to demonstrate them.

II.

The Histological Examination of the Lining of the Osseous Meatus.

As the lining of the osseous meatus is intimately connected with the periosteum, its histological relations are best examined on temporal bones that have been previously treated with Müller's fluid and subsequently decalcified, and from which the osseous meatus is cut out, along with its cuticular lining.

The cutis and the periosteum of the osseous meatus are in close connection. With the exception of that triangular space before mentioned (p. 56), which extends on the posterior superior wall, from the cartilaginous into the osseous meatus, the glandular elements are entirely wanting in the pars ossea. But the lining shows a spirally arranged ledge-shaped fold of the cutis (Kaufmann), which, on sections made vertically to the direction of the fold, shows a papillary arrangement. Besides, there are found, especially at the inner portion, in the vicinity of the tympanic membrane, vascular papillæ which, particularly in the fœtus and new-born infant, are strongly developed, and can be followed up to the peripheral part of the membrane (Moos). On injected preparations the entrance of the bloodvessels, accompanied by processes of the periosteum, can be traced from the cutis into the osseous walls of the meatus, and from here, not infrequently, into the mastoid process and the cavum tymp. These anastomoses explain the strong injection of the cutis in inflammations of the tympanic cavity and of the mastoid process.

To demonstrate the relation of the dermic layer of the meatus to the membrana tympani, sections are made on decalcified preparations, in which the meatus has been left in connection with the tympanic membrane, the cuts being carried in the long axis of the meatus, some in a frontal, others in a horizontal direction. Very instructive are those frontal sections which show the transition of the cutis and its bloodvessels from the superior wall of the meatus on to the tympanic membrane.

In like manner the osseous meatus is histologically examined in pathological cases, where it is a question of becoming acquainted with the changes in the inflammatorily infiltrated cutis, or papillæ, and polypous proliferations thereon, carcinomatous infiltration of the lining spreading from the cartilaginous meatus or from the parotid gland; the histological condition of membranous and osseous strictures and atresiæ of the meatus, and the simultaneous alterations in its osseous walls.

III.

HISTOLOGICAL EXAMINATION OF THE MEMBRANA TYMPANI.

THE tympanic membrane, loosened from the sulcus tymp. (p. 75), may be subjected to histological examination by several methods.

(a) *Histological Examination on Surface Preparations.*

This is proceeded with in various ways.

1. In the fresh state, by *the dissection of the membrane into three layers.* For this purpose two fine pincettes are used, with which, starting from the annulus tendinosus, and proceeding towards the

handle of the malleus, the dermic layer is first separated from the
radiating fibrous layer, and then the latter from the circular fibrous
layer (Toynbee, Von Tröltsch). The work of preparing is the easier,
as these layers of the tympanic membrane are only connected by a
few bundles of connective tissue. The circular fibrous layer and the
mucous membrane lamina, on the other hand, are so intimately united
that their anatomical separation is impossible. The membrana
tympani, dissected into three layers, and suspended in a small glass
vessel by means of a thread fastened to the neck of the malleus (plate,
p. 191, Fig. VI.), furnishes an instructive object for demonstration. For
microscopical examination the separated layers are cut off with a fine
pair of scissors from the manubrium, then stained and mounted after
the methods already described.

2. After removal of the handle of the malleus, *without dissecting
the tympanic membrane into its various layers.* With this object,
after the membrane has been loosened from the sulcus, the tympanic
layer covering the inner edge of the manubrium is slit length-wise
with a small sharp knife, then slightly drawn back from the lateral
surfaces of the handle, and after that the entire hammer is carefully
detached from the membrana tympani with a fine pincette. After
pencilling over both surfaces of the membrane, it is either treated
with hyperosmic acid (p. 198) or placed in Müller's fluid (p. 195). In
the former case the preparation being coloured brown by the process
of fixation, dehydrated in alcohol and cleared, shows, besides the
direction of the fibres of the substantia propria, frequently also the
bloodvessels of the cutis and mucous membrane layer, as well as the
nerve plexus of the cutis. The last-mentioned differentiate well under
gold treatment (Kessel). By fixation in Müller's fluid, the direction
of the fibres in the tympanic membrane will only stand out strongly
marked after staining with eosin, alum-carmine, or hæmatoxylin. To
examine both surfaces of the membrane, it is divided with a fine
pair of scissors into several triangular segments directed towards
the lower extremity of the manubrium; they are mounted, some
with the surface of the mucous membrane, others with that of the
cut is, upwards.

The examination of the outer side of the tympanic membrane
displays, on surface preparations behind the manubrium, after careful
brushing over of the epidermic layer, a band-like stria passing at an
acute angle to the lower end of the handle, so that there remains
between it and the manubrium a transparent space in the shape of a
triangle, directed downwards (cutis strand of Von Tröltsch, Prussak's
descending fibres). This fibrous band, radiating starlike at the lower
extremity of the handle, is the continuation of a cutaneous stria

passing from the upper wall of the meatus to the tympanic membrane, and gives passage to the bloodvessels and nerves which cross from the external meatus on to the membrane. The radiating vascular ramifications of the dermic layer, as well as the peripheral corona vascularis of the membrana tympani, and the vascular plexus behind the manubrium, appear especially clear on the membrane of the new-born infant, if the vessels be macroscopically strongly injected, and if the object be treated in a perfectly fresh state with osmic acid. Kessel asserts that he has also demonstrated the presence of a fine network of lymphatic vessels in the dermic layer of the tympanic membrane.

Although the direction of the fibres of the substantia propria, par-

FIG. 132.—Segment of the lower portion of the tympanic membrane.—h = handle of the malleus ; r = layer of radiating fibres ; c = layer of circular fibres.

ticularly that of the radiating fibrous layer, can also be examined on successful surface preparations from the dermic side, it is best demonstrated from the inner side of the membrane (Fig. 132). The radiating fibres, which arise from the tendinous ring, and in places unite at an acute angle, converge in the lower segment towards the inferior extremity of the manubrium to become attached to its spatula-shaped expansion (h). In the vicinity of this place of attachment there arises, partly through condensation of the radiating fibres meeting there, partly through deposition of small cartilage - cells, the so-called umbilical opacity of the membrana tympani (yellow spot, Trautmann). In the upper segment, the radiating fibres attach themselves to the anterior edge of the handle, and appear sharply defined at the boundary of Shrapnell's membrane. The inner circular fibrous layer

(c) does not reach up to the outer periphery of the membrane. It appears most dense in the neighbourhood of the tendinous ring, while towards the centre it diminishes in density, and is lost close to the extremity of the manubrium. It is attached, with the greater mass of its fibres in the upper segment of the membrane, to the external surface of the handle, especially at the short process.

On the inner surface of the membrana tympani there is an inconstant, greatly varying, fibrous framework (Fig. 133), which belongs to its mucous membrane, and is most strongly marked on the posterior segment. This framework, being spread over, sometimes smaller, at other times larger spaces, is interrupted by roundish-oval interstices of varying size. The processes of the framework radiate towards the

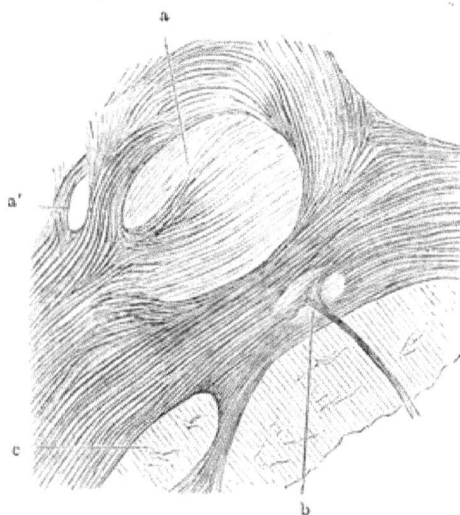

Fig. 133.—Fibrous trabecular structure in the posterior segment of the inner surface of the tympanic membrane in the adult.—a = large interspace in the trabecular structure ; b = small interspace, through which a thin process passes.

tendinous ring and the manubrium, and in part become closely united with the fibres of the substantia propria. This trabecular structure, which has been described by Gruber as a dendritic formation of the membrana tympani, but which, in my opinion, is identical with the inconstant fibrous processes in the mucous lining of the tympanic cavity may be demonstrated under a low magnifying power, even on fresh preparations, after brushing away the epithelium of the mucous layer ; but it stands out more sharply on preparations stained with carmine or hæmatoxylin, and on such as have been treated with gold or osmium.

On the tympanic membrane of the new-born infant there are found, on surface preparations of the inner side, chiefly towards the periphery, those papillæ first described by Gerlach, covered with pavement epithelium, and containing vascular loops which may be as readily found on fresh preparations as on those which have been preserved in Müller's fluid.

The vascular network of the inner surface of the tympanic membrane is best demonstrated on the new-born infant by injecting the bloodvessels from the aorta, or common carotid, with liquid glue coloured with carmine or Berlin blue (Moos). Beautiful specimens of it may also sometimes be obtained on osmium preparations, particularly if the membrane, in the fresh state, be already naturally injected. The closely meshed capillary network is developed out of an artery which runs parallel to the handle of the malleus.

The fine, non-medullated nerve-fibres of the surface of the mucous lining, described by Gerlach, may sometimes be seen on gold or osmium preparations, and form, according to Kessel, a plexus, from which nerve filaments penetrate into the epithelium. Kessel has also succeeded, by treatment with nitrate of silver, in demonstrating a system of lymphatics in the mucous covering of the tympanic membrane.

(b) *Histological Examination of the Membrana Tympani on Sections.*

The structure of the tympanic membrane, especially its relation to the manubrium, can only be represented on sections made on one preparation, at a right angle to the long axis of the handle, and on another through the entire length of the malleus and of the membrane. For these examinations it is best to use a tympanic membrane, which has been left in connection with the annulus tymp. These sections offer the advantage of representing, at the same time, the relationship of the annulus tendinosus, and of allowing the various parts of the preparation to retain their relative positions better than would be the case if the membrane were loosened out of the sulcus tymp.

As a matter of course, in all cases where the membrana tympani, together with its osseous frame and the manubrium, are to be subjected to microscopic examination, the slow decalcification of the osseous portions in picric acid, or in the nitric acid and chloride of sodium solution (p. 201), must always follow the fixing of the preparation in Müller's fluid or Vlakovic's solution (p. 195), which requires eight to ten days. Embedding in celloidin or paraffin. In embedding with celloidin, staining with alum-carmine or hæmatoxylin must be avoided, because by over-staining the embedding medium the clearness of the microscopic image suffers.

The appearance of the tympanic membrane on horizontal sections shall first be discussed.   In making serial sections in a horizontal direction through the membrane, it is best to begin with the uppermost part of the malleo-incudal articulation, and to continue the sectioning to below the extremity of the manubrium.   This mode of proceeding has the advantage of affording information as to the anatomical details of the malleo-incudal articulation and the relationship of the space between the body of the malleus and incus, and the lateral wall of the niche.   In the examination of pathological preparations this is of great importance, especially in those suppurative processes

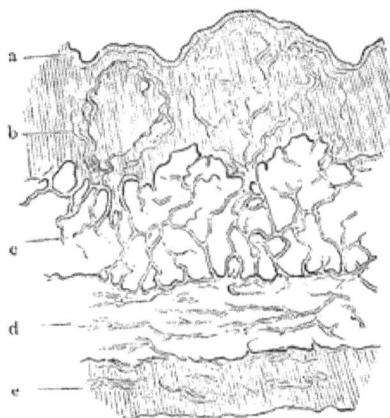

Fig. 134.—Section through the tympanic membrane of an infant.—a = epidermic layer ; b = cutis layer ; c = layer of radiating fibres, with the ramified interstitial spaces ; d = layer of circular fibres ; e = mucous layer.

which run their course in this space, and which are accompanied by perforation of Shrapnell's membrane.   Referring, as regards the anatomical relations of this region, to a future chapter, we will here first discuss the structure of the tympanic membrane on transverse sections, the relations of the annulus tendinosus to the membrane, and those of the latter to the handle of the malleus.

On microscopic sections of the tympanic membrane the following details may be distinguished from without inwards (Fig. 134) : (1) The epidermic layer, consisting of stratified pavement epithelium, with its rete Malpighii (a) composed of cylindrical cells.   (2) The connective-tissue stratum of the cutis layer, with its bloodvessels and nerves (b).   This layer appears most strongly developed at and behind the place of apposition of the manubrium to the tympanic membrane ; it here attains a thickness of 0.4 mm. (Schwalbe), and in the upper segment shows small papillary elevations.   (3) The substantia propria, having outwardly the layer of radiating fibres (c), inwardly the circular fibrous layer (d), and between the fibre bundles the interspaces and lymph channels (Tröltsch's corpuscles), which on transverse sections appear star-shaped, on longitudinal sections spindle-shaped, and have a lining of endothelial cells.   (4) The mucous membrane layer (e), consisting of connective tissue, interspersed with elastic filaments, conveying a capillary network of bloodvessels, and covered by a non-ciliated pavement epithelium.

The bloodvessels and nerves of the various layers of the membrane are examined on sections taken from preparations treated after

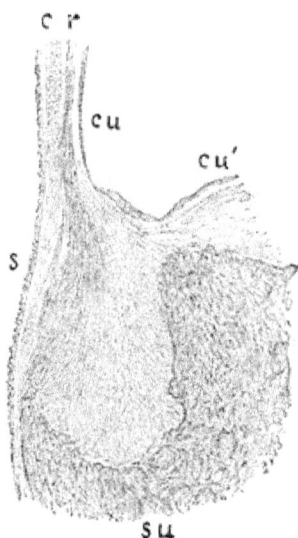

FIG. 135.—Section through the annulus tendinosus, embedded in the sulcus tymp., and through the peripheral portion of the tympanic membrane.—su = sulcus tymp., with the annulus tendinosus, the fibres of which merge, in an upward direction, into the r = layer of radiating fibres of the tympanic membrane ; cu′ = cutis of the lower wall of the external meatus, continuous with the cu = cutis layer of the membrana tympani ; c = section of the peripheral layer of radiating fibres of the tympanic membrane ; s = mucous layer of the tympanic membrane and of the tympanum with its epithelial covering.   After a preparation in my collection.

the gilding or osmic acid process; the bloodvessels alone, also on membranes where the capillary injection has been successful.   The largest transverse sections of the vessels and nerves of the cutis are seen on its thickened portion behind the manubrium.   The peripheral vascular zone shows in sections numerous anastomotic connections of vessels, passing through the interspaces of the peripheral part of the membrane to the mucosa.   On the intermediate zone, lying between the manubrium and the periphery, some capillaries run from the cutis into the substantia propria. From the nerve plexus of the cutis Kessel saw small knotty filaments penetrating into the epithelium.

Sections made through the decalcified osseous frame of the tympanic membrane are adapted for the examination of the structure of the annulus tendinosus, and its relation to the membrane on the one side, and to the cutis of the external meatus and the mucous lining of the tympanum on the other. The tendinous ring (Fig. 135) appears on sections in the form of a triangle, the serrated base of which is inserted in the rough surface of the sulcus tymp. (su). At the apex of the triangle the fibres of the annulus tendinosus pass over into the radiating fibrous layer of the tympanic membrane (r), but some of the

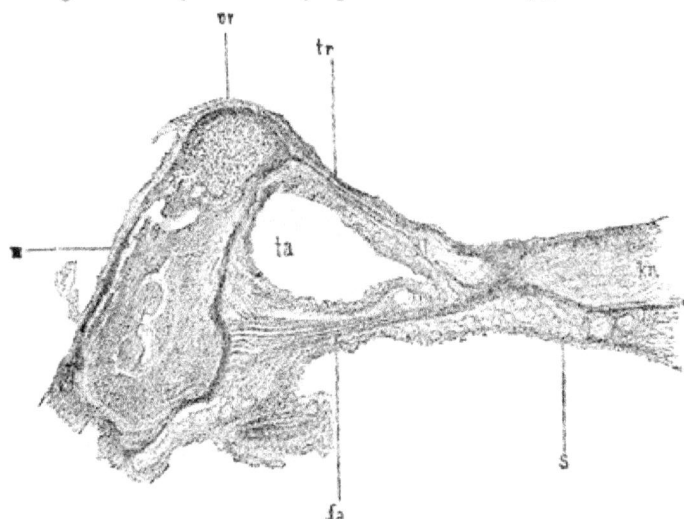

FIG. 136.—Section through the malleus and the posterior tympanic pouch at the level of the short process of the malleus.—h = malleus ; br = cartilaginous portion of its short process ; tr = posterior portion of the tympanic membrane ; fa = posterior fold of the tympanic membrane ; ta = posterior pouch of Von Tröltsch ; s = mucous lining of the tympanic cavity.

fibrous bundles of the tendinous ring may also be traced up to the periosteum of the osseous meatus, and to the mucous membrane of the cavum tymp. The texture of the ring consists of connective-tissue fibrillæ, interspersed with elastic fibres, which cross each other frequently, but contains, according to Schwalbe's and the author's investigations, no cartilage-cells.

As regards the anatomical relations of the tympanic membrane to the handle of the malleus, the uppermost sections, made through Shrapnell's membrane and the neck of the malleus, show the topographical relations between this membrane and the neck of the hammer, and, on successful preparations, the anterior and posterior boundaries of Prussak's space. For an accurate knowledge of the limits of this cavity, however, vertical sections, corresponding to the long axis of the malleus, are required, to which we shall refer again later on.

The horizontal series of sections, from the short process of the malleus to the lower spatula-shaped extremity of the manubrium, yields the following representations :

1. The section passing through the tip of the short process of the malleus and through the upper part of the posterior tympanic pouch shows (Fig. 136) the short process of the malleus (br), covered by a thick layer of cartilage-cells, which is intimately connected with the osseous substance of the malleus, and which is to be regarded, not as a peculiar cartilaginous formation of the tympanic membrane (Gruber), but as the residuum of the embryonic cartilaginous hammer.* Between the short process (br) and the osseous frame of the tympanic membrane (kn), the section of the posterior superior segment of the membrane (tr) appears covered, externally by the cutis, internally by the mucous lining of the middle ear. Medially from the membrane, the inner tympanic fold (fa) is stretched from an osseous ridge near the sulcus tymp. to the handle of the malleus, and forms, with the surface of the membrane opposite, the posterior pouch of Von Tröltsch (ta). The fibres of this fold are very similar to those of the substantia propria of the membrana tympani (Von Tröltsch). On sections of the anterior tympanic pouch, it is not in-frequently possible to follow the tense fibrous band of the ligamentum mallei ant. entering the Glaserian fissure, the chorda tymp., and, in the new-born infant, the longitudinal section of the long process of the malleus (processus Folianus). The mucous covering of the tym-panum (s) and the lining of the two tympanic pouches, which, in the adult, are thin and present only few transverse sections of vessels, appear in the new-born infant strongly tumefied, and show numerous, often very wide, bloodvessels.

2. On a section made somewhat deeper (Fig. 137), the osseous ridge (c), from which the inner tympanic fold (f) arises, appears more strongly marked, and the space of the pouch (t) larger. On sections made in this plane, we sometimes find this fold divided in the vicinity of the manubrium into two laminæ (compare p. 77), the outer of which is inserted on the posterior surface of the handle (v), while the anterior (a duplicature of the mucous membrane) passes to the

---

* Heinrich Müller (Zeitschrift für wissenschaftl. Zoologie, Bd. ix.) proved, as early as 1858, the existence of layers of cartilage-cells in the manubrium of the new-born infant and of the adult.

truncated part of the inner and posterior edge (l).   At the same time
we notice the cross-section of the chorda tymp. (ch) in its connection
with the inner tympanic fold, and sometimes also inconstant filaments
and fibrous fasciculi stretched between the membrane and the fold.

3. The somewhat deeper section of the manubrium (Fig. 138),

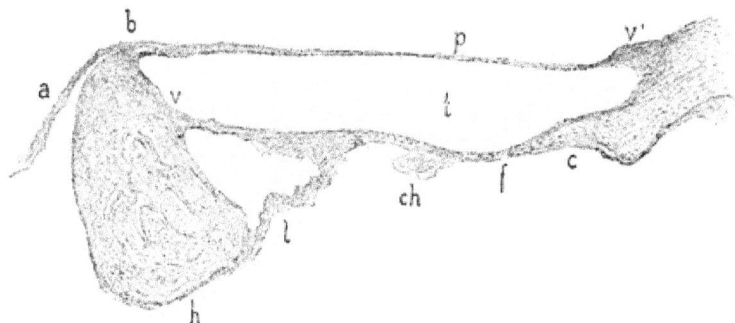

FIG. 137.—Section through the upper part of the manubrium and of the posterior pouch of
the tympanic membrane, somewhat deeper than the previous one.—h = inner surface of
the handle of the malleus ; b = its outer edge ; p = section of the posterior superior
segment of membrana tympani ; v' = its insertion in the uppermost portion of the sulcus
tymp. ; a = piece of the anterior portion of the tympanic membrane ; f = inner posterior
fold of the membrane ; c = insertion of this fold on a sharp osseous ledge, situated within
the sulcus tymp. ; v = insertion of the inner fold of the tympanic membrane at the
posterior surface of the manubrium ; l = mucous membrane lamella of the fold passing to
the inner surface of the handle ; t = posterior pouch of Von Tröltsch ; ch = section of
the chorda tymp.   After a preparation in my collection.

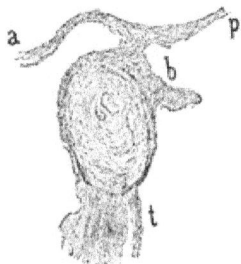

FIG. 138.—Section through the manubrium
on a level with the place of attachment of
the tendon of the tensor tymp.—a =
anterior portion of the tympanic membrane ; p = its posterior portion ; b =
slanting bridge, forming the connection
between tympanic membrane and manubrium ; t = attachment of the tendon of
the tensor tymp.   After a preparation in
my collection.

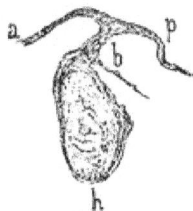

FIG. 139.—Section through the manubrium
below the previous one. — a = anterior
portion of the tympanic membrane ; p =
its posterior portion ; b = slanting bridge,
connecting tympanic membrane and manubrium ; h = posterior edge of the handle
of the malleus.   After a preparation in
my collection.

which passes through the tendon of the tensor tympani, is no longer
triangular like the sections made higher up, but shows the outlines of
an ellipsis, the long axis of which stands vertical to the plane of the

membrane. While in the region of the short process, between the handle of the malleus and the tympanic membrane, there is close contact, the section through the middle portions of the manubrium appears connected with the membrane only by a narrow bridge, which passes, not from the outer edge of the handle, but laterally from it in a slanting direction towards the posterior portion of the membrane. Similar in appearance are the next following sections of the manubrium (Fig. 139) lying below the tensor tendon.

4. On sections of the lower third of the handle, its edge is more closely in contact with the membrane, and the connecting surface appears broader (Fig. 140). The closest contact between membrane and handle exists, however, at the lower spatula-shaped extremity of

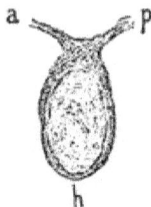

FIG. 140. — Section through the middle portion of the manubrium, below the previous one. The apposition of the handle of the malleus to the tympanic membrane is here broader and closer than in the upper portions. — a = anterior portion of the membrana tympani ; p = its posterior portion ; h = inner border of the manubrium.

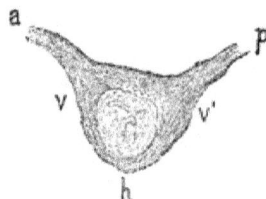

FIG. 141. — Section through the inferior spatula-shaped extremity of the handle of the malleus.—a = anterior portion of the tympanic membrane ; p = its posterior portion ; v v' = condensation of the tissue of the membrane at the lateral portions of the manubrium ; h = inner border of the latter.

the latter (Fig. 141), this flattened portion of the hammer being surrounded on all sides by the here very dense tissue of the tympanic membrane.

The mode of attachment of the radiating and circular fibres at the manubrium is, notwithstanding the investigations hitherto made, not yet quite clear. That the radiating layer is attached with the greatest mass of its fibres to the lower part of the handle, while the circular fibres are inserted at the upper portion in the region of the short process, may be taken as an established fact. The statements as to the relationship of the fibres of the substantia propria to the middle portion of the handle, on the other hand, require rectification. That the radiating fibres in the middle portion of the manubrium pass, as some suppose, everywhere crossed over to its surface, I could not quite convince myself. So far as it is possible to form an opinion from sections, I was able to see, even with a low magnifying power, that in the upper as well as the middle portions of the manubrium (compare Figs. 138 and 139), chiefly the fibres of the posterior superior segment of the membrane (p), pass over to the handle, and unite with the periosteum of its surfaces, directed towards the tympanic cavity ; that, however, the fibres of the anterior superior segment, running along the edge of the manubrium, meet the fibres of the posterior segment. Only in the lower third of the handle (Figs. 140 and 141) does there seem to be a crossing of the fibres on their passing from both sides to the edge of the manubrium.

The study of horizontal sections should be supplemented by a series of vertical sections of the tympanic membrane. Those passing through the anterior portion of the osseous frame vertically to the

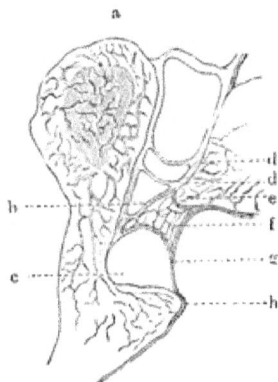

FIG. 142.—Vertical section through the malleus, Shrapnell's membrane and the malleo-incudal niche.—a = head of the malleus ; h = processus brevis mallei ; g = membrana Shrapnelli ; e = margo tympanicus ; c = Prussak's space ; b = ligamentum mallei ext. ; f = inconstant framework between the membrana Shrapnelli and the ligamentum mallei ext. ; d d = inconstant bridges of mucous membrane in the malleo-incudal niche. After a preparation in my collection.

Glaserian fissure allow, on successful preparations, the sections of the ligament. mallei ant., of the chorda tymp., of the art. tymp., and, in the new-born infant, the section of the proc. Folianus, to be distinguished.

The most important vertical sections of the tympanic membrane are those made through the entire malleus, Shrapnell's membrane, and the malleo-incudal niche. To guard against useless slanting sections, the embedded preparation must be so fastened in the microtome that the cut passes exactly through the long axis of the malleus. The transparent celloidin, in which the hammer shines through, is, for this purpose, an indispensable embedding medium. The relationship of Shrapnell's membrane to the malleus having been previously discussed (see pp. 84 and following) it need here only be observed that Prussak's space, situated above the short process of the malleus, as also the malleo-incudal niche (Fig. 142), show numerous varieties, and that the ligam. mallei extern., passing from the crista capit. mallei to the incisura Rivini, is, in the majority of preparations, inserted only on the anterior portion of the incisura; that, however, the posterior part of the ligament, below the tapering incisura (margo tymp.), unites with Shrapnell's membrane.

The methods of histological examination of the normal tympanic membrane are generally employed in pathological cases also. Whether in any special case the examination is to be made on surface preparations or on sections will always depend upon the nature of the morbid changes present. With atrophic membranes, or where cicatrices and thin calcareous deposits are found, surface preparations are preferable to sections, if the object be to represent clearly the extent of the pathological change on the surface, the partial or total absence of the fibres of the substantia propria, the disposition of the epithelium, and the course of the bloodvessels. The two methods may, however, be combined with one and the same preparation, one piece of it being used as a surface preparation, the other for sections. The latter method is especially to be recommended in cicatrix formations on the tympanic membrane, where it is to be ascertained from which of its layers the cicatricial tissue has grown, and what is the relation of the thickness of the membrane as compared with that of the cicatrix.

On the other hand, when the membrane, through inflammatory infiltration or connective-tissue proliferation on its dermic and mucous layers, through polypous hypertrophy, calcification, or ossification, has increased to several times its normal thickness, the preparations can only be histologically examined on sections. Where information is to be gained as to the relation of the pathologically altered membrane to the manubrium, it is necessary, after previous fixation, to soften the handle by slow decalcification, the same as with normal preparations (see p. 200). This also applies to markedly calcareous and partially ossified membranes. As, however, these become considerably altered through decalcification, it is advisable to examine some pieces of the morbidly changed membrane undecalcified.

In the histological examination of the various morbid forms of the tympanic membrane it is necessary, according to the results of previous research, to consider the following changes:

1. In the epidermic portion of the cutis layer, considerable proliferation and superposition of the epithelium, pigmentation of the same, formation of small bead-like globules, with epithelial (Urbantschitsch) or cholesterine contents, cone-shaped proliferations of the deeper epithelial strata into the cutis layer (Moos).

2. In the cutis layer proper: serous, purulent, and hæmorrhagic infiltration of the connective-tissue stratum, dilatation of the blood-vessels, hypertrophy of the cutis through cell-proliferation, new for-mation of connective tissue and of bloodvessels, the springing up of granulations, polypi and villous excrescences on the surface, fibrous tissue thickening of the cutis layer; finally ulceration, destruction, cicatrices, and partial atrophy.

3. In the substantia propria: serous and purulent infiltration, thickening through connective-tissue new growth, atrophy, calcareous and osseous deposits, fatty degeneration, and pigmentation.

4. On the mucous membrane layer: hyperæmia, swelling, hyper-trophy through round-cell proliferation and connective-tissue neo-plasm, polypous formation, pigmentation, perforation, and cicatriza-tion.

IV.

HISTOLOGICAL EXAMINATION OF THE LINING MEMBRANE OF THE TYMPANIC CAVITY.

THE knowledge of the normal structure of the mucous membrane of the tympanum is of great importance for forming an estimate of the pathological changes therein, because it is here that those morbid processes, which so frequently are the cause of disturbances of hearing, run their course.

The normal mucous lining of the tympanic cavity is subjected to histological examination on surface preparations and on sections.

1. *Examination on Surface Preparations.*

The most beautiful surface preparations are obtained from the promontory, from which part the mucosa may be detached in larger pieces without disturbing its continuity. This is done simply by means of a small knife rounded off at the point, with which the mucous membrane is pressed away from the inner wall of the ost. tymp. tubæ, backwards up to the vicinity of the two labyrinthine fenestræ. Particular care must be taken while manipulating in the neighbourhood of Jacobson's nerve, because here the mucous mem-brane is most intimately connected with the subjacent bone.

Far more difficult is it to obtain surface preparations from the other walls of the tympanum. One may certainly succeed in loosen-ing small portions of the membrane from the smooth malleo-incudal niche, from the posterior wall of the tympanum, the lateral osseous wall of the Eustachian tube, and from the sinus tymp. (Steinbrügge); but from the other parts of the cavity, especially from the ridgy

inferior and upper walls, where the mucous membrane either sinks
into the irregular depressions, or bridges them over, only small particles can be stripped off, which are not of much use for examination.

The normal lining membrane of the tympanic cavity in the fresh
state is little adapted for histological examination. The best surface
preparations are obtained if the entire organ of hearing, or even the
detached pyramid, be first placed in Müller's fluid, or else in Vlakovic's or Urban Pritchard's solution (p. 195), the loosening of the
membrane being only proceeded with after the preparation has been
fixed and soaked in water. By staining with eosin or alum-carmine
the various tissue-elements become so clearly differentiated that in
the microscopical examination of surface preparations which have
been mounted in Canada balsam or in glycerine, it is possible to
recognise the contours and nuclei of the epithelium, and, by deeper
focussing, the disposition of the fibrous filaments in the connective-
tissue layer, and not infrequently also the course of the larger vessels
and nerve-trunks.

In order to examine, on surface preparations, the blood- and
lymphatic vessels in the periosteal layer, the detached lining membrane must be mounted with its periosteal surface directed upwards.

The network of bloodvessels in the mucous membrane of the
tympanum is rendered visible for microscopical examination in
various ways: 1. By injecting the arterial vessels from the common
carotid, or from the aorta by means of carmine glue or Berlin blue.
But even when this is most carefully done, it is only seldom that
perfectly successful injections of the capillaries are obtained. 2. By
the action of 2% hyperosmic acid (p. 198), through which the red
corpuscles in the finest capillaries are stained brown, so that the
bloodvessels appear in the microscopical image as dark wavy lines.

If the pyramid, separated from a fresh preparation, be subjected
to osmium treatment, the dark vascular network of the lining membrane on the promontory may be recognised, even with an ordinary
magnifying lens. In addition, there are seen on the surface of the
mucous membrane minute black dots in varying numbers, which,
according to my investigations, are to be regarded as the points of
entrance of larger bloodvessels of the membrane into the osseous
wall of the promontory. For a more detailed histological examination, the lining membrane on such preparations is detached from the
bone after the manner already described, and the preparations are
mounted, some with their epithelial, others with their periosteal, side
upwards.

Less reliable for the representation of the bloodvessels of the
mucous membrane of the tympanum is the gilding method (p. 196).

But the nerve-trunks and the finer ramifications of the nerves are most beautifully shown in successful gold preparations, while the finest and most delicate nerve-plexuses, which wind round the blood-vessels of the mucous membrane of the promontory and of the tube, appear markedly differentiated in the microscopical image.

On osmic acid preparations also, not infrequently, in addition to the bloodvessels, the nerve-trunks and their ramifications in the mucous membrane are brought clearly into view. On successful osmium preparations I have often seen in the vicinity of the larger nerve-trunks ganglion cells arranged in groups, and especially frequently near Jacobson's nerve, and in the angular spaces which the branches of that nerve form with the nerve-trunk.

The examination of the pathologically altered lining membrane of the tympanum on surface preparations is also advisable where there is a slight degree of thickening of the mucosa, allowing such clearing up of the tissues as will be necessary for histological purposes. In hypertrophy of the membrane through round-cell proliferation developed in the course of chronic middle-ear suppurations, or in polypous hypertrophy, the examination on surface preparations is impossible, and the alteration of the structures can only be represented on sections. Very instructive surface-preparations are, however, obtained in those middle-ear suppurations which have run their course, where the mucous membrane is pale, smooth, and only very slightly tumefied. These, when fixed in Müller's fluid or in Vlakovic's solution, show, on the side of the epithelium, the sharp contours of the latter, here and there varicose veins and cyst-like spaces, and on the periosteal surface the varicose expansions and club-like ramifications of lymph vessels (Fig. 143), first described by me,* and, besides these, well-defined cyst-spaces lined with epithelium.

2. *Examination of the Lining Membrane of the Tympanic Cavity on Sections.*

This method makes us acquainted with the structure of the lining membrane of the tympanum in the superficial and deeper layers, and the relative position of the bloodvessels and nerves therein. The lining membrane, when detached from the bone, not being suitable for sections, on account of its thinness, it is advisable to cut it in connection with the decalcified bone. This has the advantage that the mucous membrane is retained *in situ* on the section by the decalcified bone, and the bloodvessels of the membrane in their relationship to those of the adjoining osseous wall, and any consecutive pathological changes in the bone itself, are represented simultaneously.

* Lehrbuch d. Ohrenheilk. 2. Aufl. S. 287.

The fixing of the tissue-elements, the decalcification and subsequent microscopico-technical procedures are the same as before
described.   The bloodvessels and nerves are examined on preparations treated with hyperosmic acid and chloride of gold.

The lining membrane of the tympanum shows, on sections of the
different walls of the latter, a varying degree of thickness.   The
epithelium of the membrane on the promontory (Fig. 144) is a
simple pavement epithelium, but on the other walls, according to the
investigations of Kölliker, a ciliated columnar epithelium, which,
according to Brunner, shows tapering basal cells; according to Kessel,
also goblet cells.   The connective-tissue stratum consists of fibrillar,

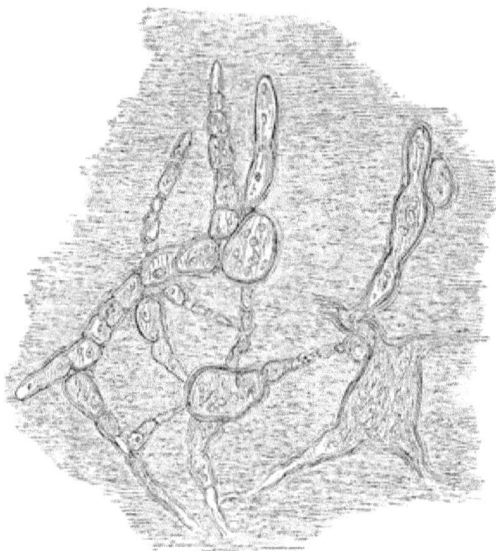

FIG. 143.—Dilated network of lymphatic vessels in the deeper layers of the covering of the
promontory in a phthisical individual, 27 years old, who had otorrhœa since childhood.
(Right ear.)   Hartnack Obj. 7.

reticulated filaments, while in the layer corresponding to the periosteum the dense prolongations of connective tissue are placed parallel.

On microscopic sections of the mucous membrane of the tympanum in decalcified preparations, which have been injected or
treated with hyperosmic acid, it is easy to determine the relationship
of the bloodvessels in the superficial and deeper layers of the mucosa.
The superficial vascular network reaches as far as the epithelial layer,
and in the new-born infant vascular loops are seen to enter the villi
(Moos) and papillæ (Lucæ).

The bloodvessels in the deeper layers, accompanied by the connective-tissue processes of the periosteal layer (Fig. 144), penetrate

into the bone covered by the membrane, to become connected with the bloodvessels of the same. On some preparations I have also found bloodvessels passing in a spiral course from the surface of the mucosa into these fissures in the bone (c'). On the promontory it is possible, as I was the first to point out,* to demonstrate, on successful osmic acid preparations, anastomotic communications of the bloodvessels in the mucous membrane of the middle ear with those of the lining of the labyrinth, by means of vessels running in the osseous wall.

Glandular elements are found in greatly varying numbers only in the anterior portion of the tympanum, corresponding to the ost. tymp. tubæ. In the middle and posterior part of the cavity they are wanting, or occur but seldom, and then only isolated. Wendt, by boiling the membrane in acetic acid, has demonstrated the presence of mucous glands on the promontory.

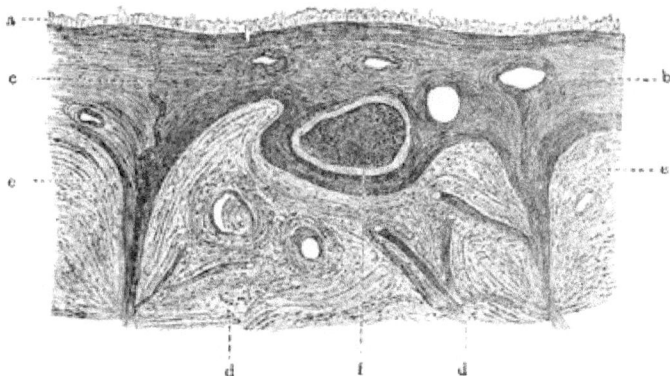

Fig. 144.—Section through the mucous membrane of the promontory (decalcified osmic acid preparation).—a = epithelium ; b = section of a bloodvessel in the connective-tissue layer, a branch of which penetrates into the funnel-shaped depression of the bone ; c = bloodvessel, passing from the surface of the lining into the bone ; d d = osseous wall ; e e' = funnel-shaped depressions in the bone, where the periosteal layer of the mucous lining enters ; f = section of a large nerve trunk in the connective-tissue layer with the mucous lining.

The already-mentioned (pp. 165 and 112) inconstant folds and filaments stretched in the tympanum consist of a connective-tissue base, which is covered with an epithelium, and the fibrous bundles of which radiate in the form of a fan. By the side of and between these folds, knotty filaments are stretched both in the tympanic cavity and in the antrum mast., which, as I was the first to demonstrate† (Fig. 145), prove in the microscopical examination to be oval or angular stalked formations with a stratified, bulb-like structure. The stalk (a),

* Wiener med. Wochenschrift, 1876, u. Arch. f. Ohrenheilk., Bd. XI.
† Ibid., 1869, 20 Nov., u. Arch. Ohrenheilk., Bd. V.

entering the rounded extremity, passes sharply defined through the structure, and emerges at the opposite end, to become attached to the prolongation of a fold, or to the osseous wall. For the examination of these small bodies, which may attain a size of 0.1—0.9 mm., organs are suitable in which, after removal of the tegm. tymp. et mast., the folds of mucous membrane and filaments alluded to will be found in considerable numbers. If this frequently web-like spreading network be loosened with a fine pincette, then removed, stained with eosin or alum-carmine and placed on the slide, a number of these stalked formations will commonly be discovered between its fan-shaped fibrous processes.

For the study of the pathologico-anatomical changes in the lining membrane of the tympanic cavity, the examination on sections is indispensable. This is especially the case in hypertrophies of the membrane caused by chronic catarrhs and suppurative processes, where sections alone will enable us to examine the changes of the epithelium, the superposition of its layers, the degree of round-cell

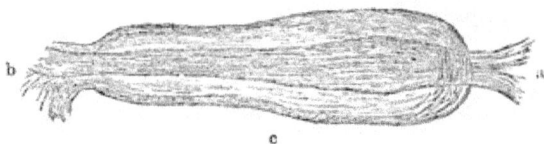

FIG. 145.—Pedunculated formation from the tympanic cavity of an adult.

and connective-tissue neoplasms, the presence of papillary and polypous excrescences on the surface of the lining and cystic spaces therein, the changes in its blood- and lymphatic vessels, and in the osseous wall covered by the membrane. But even with a not markedly hypertrophic membrane—*e.g.*, in some forms of middle-ear sclerosis—the examination on section preparations cannot be dispensed with, when it is a question of ascertaining the finer structural changes in the mucous lining.

THE HISTOLOGICAL EXAMINATION OF THE PATHOLOGICAL CHANGES IN THE TWO LABYRINTHINE FENESTRÆ.

The technical proceeding in the histological examination of the morbid changes in the two fenestræ and in their niches requires, considering the physiological importance of this region of the tympanum, a more detailed discussion.

First, as regards the changes in the fenestra ovalis and its niche, the direction of the cut will have to vary according to whether the pathological alterations in the niche alone, or, at the same time, also those at the ligamentum orbiculare stapedis, are to be brought under examination. In the first case, the cuts are made parallel to the inner tympanic wall through the decalcified pyramid, previously reduced down to the capsule of the labyrinth. Such sections (Fig. 146) show the circumference of the niche of the fenestra ovalis, its mucous lining, and the relationship of the semilunar sections of the crura of the stapes. Sections made in this direction through the normal ear show the disposition of the connective-tissue bridges and filaments, stretched in varying numbers between the crura of the stapes and the wall of the niche. These narrow bridges form, on the setting in of an inflammatory process, a predisposing cause for the commencement of anchylosis of the crura of the stapes. In patho-

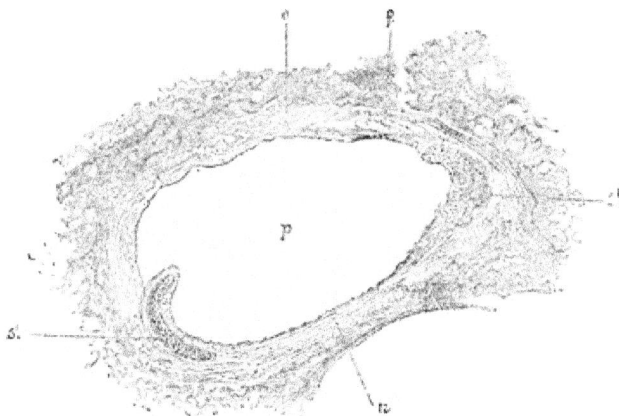

Fig. 146.—Anchylosis of the crura of the stapes with the lower wall of the pelvis ovalis, from the left ear of a woman 48 years of age, who, while suffering from constant tinnitus, had gradually become deaf (20 years on the left, 10 years on the right side), and who died in one of the charities. Microscopic section through the pelvis ovalis and the crura of the stapes.—p = pelvis ovalis ; o = upper wall of the niche of the fenestra ; n = mucous lining of the lower wall of the niche, which has become changed into sclerosed tissue ; st st = sections of the anchylosed crura of the stapes, enveloped in dense sclerosed connective tissue.

logical investigations such parallel sections prove especially instructive for the study of anchylosis of the crura of the stapes with the lower wall of the niche (Fig. 146), and for the purpose of showing the relationship of the crura in those excessive connective-tissue proliferations which, wholly or in part, fill up the niche, and envelop the stapes on all sides.*

A second mode of sectioning, in order to represent the normal and pathological relations in the fenestra ovalis, consists in carrying horizontal cuts parallel to the two crura of the stapes in such a way

* Compare A. Politzer: Sur les changements pathologiques dans la fenêtre ovale et dans la fenêtre ronde, qui se produisent chez les affections de l'oreille moyenne. Comptes rendus du IVme Congrès otologique à Bruxelles, 1888. Publié par le Dr. Ch. Delstanche, 1889.

that several cuts meet the greatest longitudinal diameter of the foot-plate, the two crura, and the capitulum (Fig. 147). These sections offer the advantage that not only the connective-tissue and osseous adhesions occurring between the crura and the walls of the niche become clearly visible, but also the histological changes in the fibrous ligament of the oval window, and in the cartilaginous covering of the stapedio-vestibular articulation. The number of serviceable sections obtainable by this method is, however, limited to two or three at most, because all those made above and below the greatest diameter of the fenestra must meet the stapedio-vestibular articulation in a slanting direction ; and, if the cut pass beyond the margin of the fenestra ovalis or of the stapes, they will prove of little use for examination.

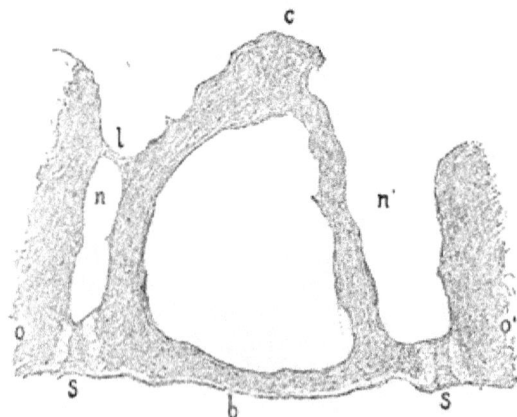

Fig. 147.—Horizontal section through the stapes and the niche of the fenestra ovalis.— c = capitulum stapedis ; b = basis stapedis ; o o′ = section of the border of the fenestra ovalis, covered with cartilage ; n n′ = niche of the fenestra ovalis ; s s = section of the ligamentum orbiculare stapedis ; l = normal mucous-membrane bridge between the niche of the fenestra ovalis and the crus of the stapes. After a preparation in my collection.

The direction of the cuts here described corresponds to the course of that piece of the facial nerve which passes above the fenestra ovalis. If, therefore, they are to meet both the capitulum and the two crura, the preparation should be so embedded that the knife of the microtome may move parallel to that piece of the nerve.

A third mode of making microscopical sections through the labyrinthine fenestræ is the frontal. It offers the advantage of allowing the changes in the oval and round windows to be examined at the same time (Fig. 148).

For making frontal sections, the preparation must be so fastened in the microtome that the posterior extremity of the pyramid is directed upwards. If it be intended to examine the two fenestræ, together with the labyrinth, the cuts carried in this direction will first bring the transverse sections of the semicircular canals into view. Should these show nothing abnormal, the entire portion of the pyramid lying behind the fenestræ may, to save time, be removed, and the sectioning through the fenestræ at once proceeded with.

As the frontal cuts fall perpendicularly to the long axis of the oval window, a greater number of useful serial sections for the examination of the histological changes in the stapedio-vestibular articulation is obtained than by carrying the cuts horizontally. On the other hand, the more exact anatomical relations of the crura of

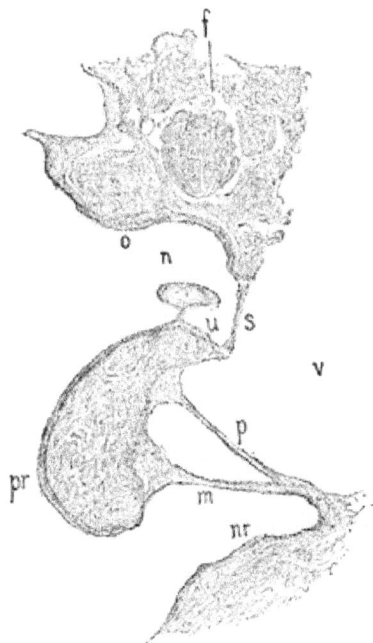

Fig. 148.—Frontal section of the inner tympanic wall through the two labyrinthine fenestræ. —s = basis stapedis ; n = niche of the fenestra ovalis with a portion of the crus of the stapes, which has been divided obliquely ; o = upper, u = lower wall of the niche ; pr = section of the promontory with its mucous covering : m = membr. fenestræ rotundæ ; nr = niche of the fenestra rotunda ; p = lamina spiralis secundaria ; f = section of the n. facialis ; v = vestibule. After a preparation in my collection.

the stapes to the walls of the niche will not be visible on frontal sections.

The anatomical relations of the membrane of the fenestra rotunda, and the normal and pathological formations occurring in its niche, present themselves most clearly in sections of this kind. Here we see in the normal organ of hearing (Fig. 148), the insertion of the membrana fenestræ rotundæ (m) in the groove, its thickness and structure, the mucous covering of its outer surface, its relationship to the lamina spiral. secundaria (p), and sometimes, in the niche of the fenestra (nr), a trabecular structure of connective tissue stretched in exceedingly varying form between the walls of the niche. It is evident that, in acute inflammations of the lining membrane of the middle ear, the exudation secreted in the niche of the round window is retained in the spaces of this structure, and that pressure upon the membrane must cause disturbance of hearing to a high degree. There is also no doubt that in chronic inflammatory processes, particularly in middle-ear suppurations, the normal folds and bridges in this niche supply the foundation of excessive round-cell and connective-tissue prolifera-tions, which fill up the niche and adhere to the membrane of the fenestra. The formation of such connective-tissue plugs, which block the niche, may, however, also proceed from its inflamed mucous membrane.

The frontal cuts through the two labyrinthine fenestræ, moreover, offer a special advantage, as they yield a perfect section of the pro-montorial wall (Fig. 148, pr), and of its mucous lining, as well as of the covering of the labyrinth, which will enable us to examine the morbid changes occurring there, and their relations to the vessels of the osseous wall of the promontory. In addition to this, such cuts render it possible to examine any alterations in the section of the facial nerve (f), in the vestibule (v), and in the commencing portion of the cochlea. In pathological cases where, simultaneously with the changes in the niches of the fenestræ, the morbid processes developing in the labyrinth are also to be investigated, it is advisable to cut the entire pyramid, from the anterior end of the cochlea to the posterior extremity of the horizontal semicircular canal, into serial sections, on which, in addition to the pathological appearance in the mucous membrane of the inner tympanic wall, the structural changes in the cochlea, vestibule, semicircular canals, and in the trunk of the acoustic nerve, can be represented.

V.

The Histological Examination of the Ossicula.

For the histological examination of the ossicula and their articular connections, only decalcified preparations can be employed.

1. *Malleo-Incudal Articulation.*

To obtain serial sections of the malleo-incudal articulation, after dividing the tensor tendon and the incudo-stapedial joint, the pars tymp., with the tympanic membrane and the malleo-incudal connection, are separated from the pyramid and fixed in one of the before-enumerated solutions (p. 195), or in hyperosmic acid, and slowly decalcified (picric acid or nitric acid). The preparation, embedded in celloidin, is cut perpendicularly to the long axis of the malleus and incus—that is, parallel to the rotation axis of the malleo-incudal joint. The objects, fixed in the chromic acid solutions, are subjected to double staining in alum-carmine or eosin, or in eosin and hæmatoxylin. Osmic acid preparations require no further staining. The same method is also employed in the examination of the other articular connections of the ossicles.

In the horizontal section of the malleo-incudal joint (Fig. 149) the two small bones are connected by means of a capsular ligament (c), which is attached to the depressed margins of the articular surfaces, and permits ample mutual displacement of the bones. A connective-

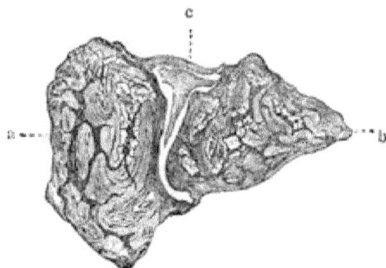

FIG. 149.—Section through the malleo-incudal articulation.—a = malleus ; b = incus ; c = capsular ligament with the wedge-shaped meniscus. Preparation treated with hyperosmic acid.

tissue cartilage, first observed by Pappenheim (*l.c.*), and accurately described by Rüdinger, is seen projecting, in the form of a wedge-shaped meniscus, from the inner wall of the capsule into the cavity of the joint, and adapting itself closely to the cartilaginous surfaces of the malleus and incus.

2. *Incudo-Stapedial Articulation.*

On a decalcified organ, after removal of the tegmen tymp., the tendon of the m. stapedius, the crura of the stapes, and the long process of the incus are divided with a small pair of scissors, and the incudo-stapedial connection, thus loosened, is lifted out. The embedding and cutting (horizontal) require great care, on account of the smallness of the object. Successful sections of this joint are occasionally obtained when sectioning the entire organ in horizontal series.

The incudo-stapedial joint (Fig. 150) is formed by the concave

articular surface of the capitulum stapedis (g) and the corresponding
convex surface of the ossicul. lenticulare Sylvii (o). Both surfaces are
covered with a thin layer of hyaline cartilage. The fibrous capsular
ligament, which connects the extremities of the joint, allows ample
lateral displacement of the same. The articulation is not an amphi-
arthrosis (Brunner), but a true joint, provided with a cavity (Eisell),
the hollow of which is divided into two clefts by a meniscus of fibrous
tissue (Rüdinger).

3. Stapedio-Vestibular Connection.

In order to examine the normal relations of this joint, the
separated pyramid, after previous fixing in chromic acid solutions or

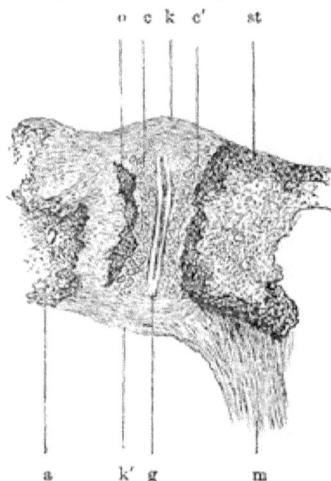

Fig. 150.—Section through the incudo-stapedial articulation.—a = terminal piece of the
long crus of the incus, and, connected with it by fibrous tissue o = ossicul. lentic. Sylvii ;
st = capitulum stapedis ; g = articular fossa with the meniscus ; c c′ = hyaline cartilage
covering of the articular surfaces ; k k′ = articular capsule ; m = tendon of the musc.
stapedis.

in hyperosmic acid, is slowly decalcified and embedded in celloidin or
paraffin. The celloidin embedding has the advantage of allowing the
stapes to shine through, so that the direction of the horizontal or
frontal cut (p. 233) through the fenestra ovalis can be accurately
determined. This is very important, as slanting sections, which give
indistinct images, are to be avoided.

The margin of the fenestra ovalis (Fig. 151, a) and the foot-plate of
the stapes (b) are covered by a layer of hyaline cartilage (Toynbee,
Magnus), which is continuous over the whole of the foot-plate (Eisell).
This cartilage layer is to be regarded as the residue of that part of the
foot-plate which is derived from the labyrinthine capsule (Gradenigo).

According to my investigations, this layer of cartilage is very thin on the vestibular surface of the stapes—at times, however, so thick that in some places it reaches as far as the tympanal side of the stapes. At the margin of the fenestra ovalis it sometimes descends deeply into the osseous tissue, and occasionally I have seen it extending outwards towards the wall of the niche—*i.e.*, to a spot which lies outside the stapedio-vestibular connection.

The fenestra ovalis and the edge of the foot-plate are connected through a fibrous elastic ligament (c), the radially disposed fibres of which are intimately united with the perichondrium.

The histological examination of the articular connections of the ossicula in pathological cases is made according to the same methods observed with normal preparations. Considering the physiological importance of the ossicular chain, it is hardly necessary to enter more fully into details to show the value of these examinations for the

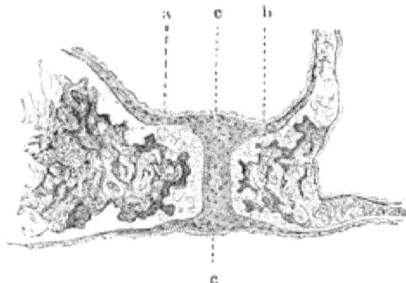

FIG. 151.—Section through the stapedio-vestibular connection.—a = border of the fenestra ovalis, covered with a layer of cartilage ; b = edge of the foot-plate of the stapes, covered with cartilage ; c c = section of the ligament. orbic. stapedis.

pathology of the organ of hearing. In anchylosis of the malleo-incudal articulation, in caries of the malleus and incus, with connective-tissue masses in the upper tympanic space, by which these bones are enveloped, the microscopic sectioning of the joint must be effected in a horizontal direction. In anchylosis of the hammer-head with the tegmen tymp., again, in all cases where the relations of the malleus and incus to Shrapnell's membrane, and to pathological processes occurring in the malleo-incudal niche, are to be clearly shown, sectioning in a frontal (vertical) direction will have to be decided upon. The incudo-stapedial articulation is always cut horizontally. Of the pathologico-histological changes in this part little is as yet known. The cuts through the stapedio-vestibular connection are carried in a horizontal or frontal direction (p. 234), in case the relationship of the crura of the stapes to any simultaneous processes in the pelvis ovalis is to be taken into consideration : in the microscopical examination attention must be directed to the structural changes in the cartilage layer (calcareous incrustations, atrophy, fusion), in the ligam. orbiculare staped. (shrinking, calcification, ossification, destruction), and at the foot-plate (thickening of the periosteum of the surface of the labyrinth, hyperostosis, wearing away, etc.).

VI.

HISTOLOGICAL EXAMINATION OF THE INTRA-TYMPANIC MUSCLES OF THE EAR.

THE disposition of the muscular fasciculi in the osseous canals enclosing them (p. 98), and the relations of the two intra-tympanic muscles to the tendons, are examined on decalcified organs, both on horizontal and transverse sections.

1. *Musculus Tensor Tympani.*

Microscopical sections in the long direction of this muscle are made by carrying a series of cuts horizontally through the entire decalcified organ. In order to obtain specimens which, in addition to the muscle, show also the longitudinal section of the tensor tendon and its attachment to the handle of the malleus, the tegmen tymp. is so far removed as to allow the tendon to become visible; it will then be possible to fasten the embedded preparation in the microtome in such a way that the knife passes simultaneously through muscle, tendon, and manubrium. On transverse sections of the muscle, obtained by sectioning the entire organ in frontal series, it is possible to demonstrate the varying form of the outline of the muscle, and the marked tapering of its anterior extremity springing from the cartilaginous tube.

The muscular fasciculi of the tens. tymp. arise, according to Helmholtz, from the periosteum of the upper surface of the canal; the tendon of the muscle can be followed for some distance into the canal.

2. *Musculus Stapedius.*

On the pyramid, which after separation from the pars tymp. has been decalcified, the muscle is sectioned longitudinally corresponding to the direction, elongated backwards and downwards, of its tendon. Successful sections show the pyriform shape of the pennate muscle, the exit of its tendon at the apex of the eminentia pyramidalis, and the insertion of the tendon below the capitul. stapedis. On transverse sections the muscle appears, according to my examinations, triangularly prismatic. The statement of Zuckerkandl, that between the muscular bundles of the m. tensor tymp. and

m. stapedius, adipose tissue is here and there deposited, I can confirm from my own preparations.

Transverse sections of the m. stapedius meet also the descending portion of the nerv. facialis, which allows the topographical relations of these two to be represented. In the new-born infant one finds a more or less constant communication between the eminentia stapedii and the canal. facialis, so that the muscle and the nerve come into immediate contact ; in the adult the separation of the two spaces is more complete, but here, too, one very often meets gaps of various sizes on the posterior portion of the eminentia stapedii* adjoining the facial canal.

The examination of the intra-tympanic muscles in pathological cases has already been discussed (p. 100).

VII.

HISTOLOGICAL EXAMINATION OF THE EUSTACHIAN TUBE.

THE structure of the Eustachian tube, the relations of its cartilaginous and membranous portions, as regards form, the position of the mucous glands and their discharge into the canal, the mode of insertion of the tubal muscles, are most comprehensively illustrated by sections made transversely to the long axis of its canal. Longitudinal sections are only made use of to represent microscopically the union between the cartilaginous and osseous portions.

For the purpose of making transverse sections of the cartilagino-membranous portion, the tube is prepared out, together with the muscles inserted therein, according to the method described on p. 104. Fixing, hardening, and embedding are effected after the general methods, details of which have been given before (p. 194). Where the relationship of the Eustachian tube to the basilar fibro-cartilage and to the neighbouring vessels and nerves is also to be examined, the tube, with the corresponding piece of the basis cranii, is sawn out, and, after sufficient fixing, decalcified in chromic acid solutions. Best adapted for this purpose, because less bulky and more quickly decalcified, are organs of new-born infants.

On transverse sections of the cartilagino-membranous tube (Fig. 152) the medial cartilaginous plate (a) appears bent over at the upper edge in the form of a hook (b).

Close to the insertion of the cartilagino-membranous part in the osseous portion of the tube, the medial cartilaginous plate is narrower than where it is bent laterally. Lower down, however, the breadth of this plate rapidly increases, while the lateral part (b), which

---

* Compare A. Politzer, Ueber das Verhältniss des Musc. stapedius zum Nerv. facialis. Arch. f. Ohrenheilk., Bd. ix.

is turned in, forms, along the whole tube, a narrow strip, which roofs over the tubal canal.

At the convoluted extremity of the cartilage hook is inserted the pars membranacea of the Eustachian tube, which passes below into the fascia salpingo-pharyngea, forming, with the cartilage hook, the lateral wall (i) and the floor (d) of the tube. In the vicinity of the hook the pars membranacea is thinner than in the lower portion, and its thickness increases from the ost. pharyng. tubæ towards the upper portions, being strengthened by the addition of fibrous tissue from the spina angularis of the sphenoid bone, and by a small rod of cartilage pushed into this tissue (Zuckerkandl).

The relations of the space (c), formed by the cartilage hook (b) below the spot where it is twisted, vary in the several sections of

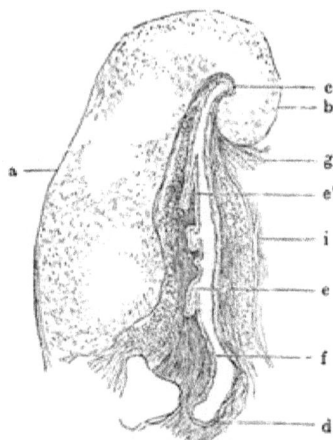

FIG. 152.—Transverse section through the cartilaginous Eustachian tube.—a = medial cartilage plate ; i = membranous plate ; b = cartilage hook ; c = space below the same ; d = floor of the Eustachian tube ; e e′ = folds of mucous membrane ; f = cylindrical epithelium ; g = musc. tensor palat. mollis ; h = musc. levator palat. mollis.

the tube. In the portions lying near the osseous part of the tube, there is no contact of the walls of this small space, while in the middle portion the contact of the medial and lateral walls is complete.

The mucous membrane, which has numerous folds, especially in the lower part of the tube, shows a ciliated cylindrical epithelium, with basal and goblet cells (Schultze). The greater mass of the strongly developed acinose mucous glands lies in the mucous membrane covering the medial cartilaginous plate. On successful sections, the excretory ducts of these glands may be followed as far as their lateral discharge into the tubal canal. The glandular layer reaches up to the perichondrium; not infrequently, however, single glands extend through the inconstant fissures of the tubal cartilage, as far as the connective tissue outside the Eustachian tube.   In addition to the mucous glands there are found, from the ost. pharyng. tube to the osseous Eustachian tube, numerous follicular glands (Gerlach), and in the lower portion of the tube in the adult scanty adenoid tissue. Finally, the transverse sections here sketched throw light on the relationship of the tubal muscles to the cartilagino-membranous part of the tube.   The bundles of the tensor palat. mollis (g) (abduct tubæ, Von Tröltsch) arising from the cartilage hook, correspond in their course with the longitudinal direction of the muscle, while those of the m. levator palati mollis (h), passing along the floor of the membranous tube, appear in the transverse section.

Transverse sections of the osseous part of the Eustachian tube show the outline of the canal to be, anteriorly, that of an irregular triangle; posteriorly, that of a quadrangle. (Compare pp. 39 and following.)   The mucous membrane contains glands, though fewer than the cartilaginous portion.   At the ost. tymp. tubæ we find on the medial and lower walls papillary elevations of the mucous membrane.

The structural changes in the Eustachian tube in pathological cases are also examined on transverse sections.   The most striking alterations in the tube are met with in chronic diffuse catarrhs, and chronic suppurations of the middle ear.   The epithelium is frequently cast off, or transformed into one of several layers, permeated with fat-globules and pigment.   The mucous glands are either hypertrophic, the acini and excretory ducts dilated, or wasted by atrophy.   The submucous connective tissue is either traversed by round cells, or thickened, condensed, and shrunken by new formations of connective tissue, whereby the tubal canal may be narrowed in various degrees. Adhesions in the canal are of rare occurrence : most frequently they are caused in chronic middle-ear suppurations, through the formation of granulations at the ost. tymp. tubæ.   The tubal cartilage shows, in pathological cases, either a normal state, or atrophy, fatty degeneration, calcareous deposits, and pigmentation.

Addenda.—For the histological examination of the processus mastoid. in the normal organ of hearing, frontal sections of decalcified preparations are best adapted.   These sections show the outlines of the antrum mastoid., the shape and arrangement of the pneumatic cell-spaces of the mastoid process, and their relative position to the antrum, the section of the very delicate mucous-membrane lining of the latter, and of the pneumatic cells, and, not infrequently, of the ramified trabecular structure of connective tissue, stretched out in the antrum, with the already-described (p. 232) pedunculated and

16—2

stratified corpuscles. Besides, on sections of partly diploëtic mastoid processes (p. 106), the characteristic differences between pneumatic cells and the diploëtic spaces can easily be determined. If the cut through the mastoid process meets also the sinus transversus, not infrequently the openings of the veins of the mastoid, passing along the osseous wall of the sinus into the sinus transversus, may be observed.

In pathological cases, microscopical sections through the decalcified mastoid process are prepared chiefly when the pneumatic spaces are filled up with new-formation of connective tissue, in eburnation of the mastoid process, and in cholesteatomatous formation therein, exceptionally only, in carious or necrotic processes. It will depend on where the seat and extent of the pathological change in the mastoid process may be found, whether, in a given case, the section is to be made horizontally or frontally.

VIII.

HISTOLOGICAL EXAMINATION OF THE LABYRINTH.

THE histological examination of the membranous labyrinth is one of the most difficult tasks. This will be understood if we consider the complicated relations of the membranous labyrinth, the peculiar way in which it is fixed in the osseous capsule, and the extreme delicacy of the terminal organs of the acoustic nerve apparatus, which are displaced on the slightest mechanical interference.

Notwithstanding the unusual interest which, since the epoch-making discovery of Corti, prominent anatomists such as Kölliker, Max Schultze, Deiters, Böttcher, Waldeyer, Gottstein, Hasse, Löwenberg, Ranvier, Retzius, Kuhn, Urban Pritchard, Winiwarter, Nuel and others have directed to the investigation of the histological relations of the labyrinth, the microscopical dissection of its structures remained up to recent times imperfect, and it is only of late years, especially through the perfection of the methods of decalcifying and embedding, that material progress has been made in the further development of the normal and pathological histology of the labyrinth.

For the study of the histology of the normal labyrinth, the majority of the above-mentioned anatomists used organs of newly killed animals, most frequently of guinea-pigs and cats. Indeed, circumstances are less favourable for obtaining perfect microscopical preparations of the labyrinth in the human subject than in animals, where, immediately after death, the organ can be placed in one of those chemical fluids which effect differentiation of the tissue-elements of the labyrinth only, when the objects are perfectly fresh.

In the human subject, on the contrary, where the post-mortem examination is, by law, not allowed to be undertaken until twenty-four hours after death, the labyrinth undergoes, in consequence of decomposition setting in, changes which interfere with the histological examination. To this must be added that in chronic disease of any organ, or in general disease ending fatally, the tissue changes developing in the whole organism must also affect the auditory apparatus, and alter the tissue-elements of the labyrinth to such a degree as to render the histological examination very difficult. This fact must be reckoned with when examining pathological labyrinths, if errors respecting their condition are to be avoided.

To obtain serviceable preparations of the human labyrinth, organs of younger individuals should be selected where death ensued after acute diseases, which had run their course rapidly, or where it occurred suddenly through apoplexy, paralysis of the heart, or through suicide, etc. As, however, the microscopic preparation of the labyrinth requires considerable experience, it is advisable to practise by making microscopical sections of the labyrinths of young guinea-pigs and cats.

The most beautiful microscopic representations of the normal membranous labyrinth are obtained by the treatment of fresh organs with hyperosmic acid (p. 198), with the solution of Tafani (p. 195), or with chloride of gold (p. 196). The latter is chiefly employed in the preparation of Corti's organ, for the purpose of showing the connections of the nerve-filaments of the acusticus with the hair-cells. While referring, for the respective details of the work of preparing to the methods previously described, we will, in addition, give a method employed by Dr. Katz, of Berlin, for the representation of Corti's organ, which, to judge from the preparations kindly sent to me, yields excellent results.

The preparation, in as fresh a state as possible, is first placed in a $\frac{1}{4}\%$ solution of osmic acid (30 ccm. of fluid), and left in it for about ten hours. To this liquid is to be added four times the quantity of a mixture of chromic and acetic acid ($\frac{1}{2}\%$ chromic acid, and 1% glacial acetic acid). After three to four days, only the quantity mentioned of the chromic and acetic acid solution is renewed, and the preparation again left therein for about four days. Next follows the decalcification of the petrous bone in a 3 to 5% solution of nitric acid, according to the size and consistence of the object. After complete decalcification, the bone is, for the purpose of hardening, dehydrated in alcohol, and embedded in celloidin.

The embedding is the more successful if the superior semicircular canal be opened with a file, and the preparation placed first in the thin, and then in the thick-fluid celloidin solution (see 'Celloidin Embedding,' p. 203).

The addition of the acetic acid in this proceeding is very important, as it causes the chromic acid to penetrate into the tissue better and more quickly. With very dense human petrous bones, Katz employs, for the purpose of decalcification, hydrochloric acid and chloride of palladium (Waldeyer), which is frequently more effective and rapid in its action.

Regular fixing of the tissue-elements, slow and careful decalcification, and complete filling up of the labyrinthine cavity with celloidin, are, therefore, the principal conditions for obtaining successful sections of the labyrinth.

The sections intended for the examination of the labyrinthine structures are made in a horizontal or frontal direction; for the study of the anatomical and topographical relations of the membranous labyrinth both methods of sectioning are indispensable.

1. Histological Examination of the Structures of the Vestibule on Horizontal and Frontal Sections.

Series of horizontal and frontal sections of the pyramid in the new-born infant will give a general view of the topographical position of the two vestibular saccules and of the ampullæ. The horizontal sections are made parallel to the upper surface of the pyramid. The uppermost portion of the vestibule shows the roundish section of the sinus utriculi super., which is, at its entire circumference, in connection with the vestibular wall by means of a finely-meshed connective tissue. On sections lying somewhat deeper there appears, behind the utriculus, the section of the ampulla superior with its crista ampullaris, and if the cut has been in the right direction, also the discharge of this ampulla into the utriculus. In a still lower section passing above the fenestra ovalis (Fig. 153), the utriculus (s), into which the ampulla extern. (a) discharges, shows an oval form; its diameter appears here smaller than that of the upper part of the sacculus (r), called by Retzius sinus utricularis sacculi. Here, too, the saccules are nearly everywhere attached to the wall of the vestibule by a delicate network of connective tissue, the interstices of which are filled with perilymphatic fluid. Where the anterior wall of the utricle is in direct contact with the posterior wall of the saccule (septum vestibuli), we find the 2—4 mm. broad macula acustica utriculi (na), with which the utricular branch (n) of the ramus vestibuli of the acoustic nerve becomes connected, and the more detailed description of which remains to be given. The macula acustica sacculi (na'), lying on the anterior wall of the saccule, to which a separate nerve branch (n') of the ramus vestibuli passes, is narrower (1.5 mm. Schwalbe). According to my examinations,* there exist anastomotic communications between the ramulus utriculi et sacculi.

On the deeper sections, falling in the region of the fenestra ovalis and below it, the diameter of the utricle appears greater than that of the saccule, which, diminishing downwards, passes over into the canalis reuniens Hensenii.

The utricle is attached, in and below the projection of the fenestra

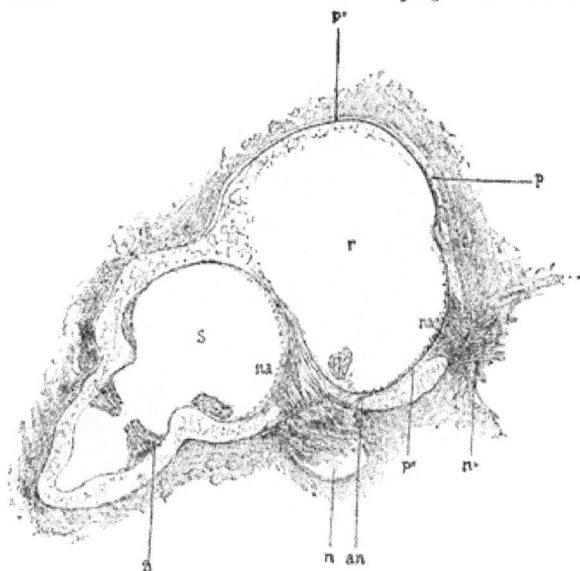

FIG. 153.—Horizontal section through the vestibule above the fenestra ovalis of a new-born infant.—s = utriculus ; a = ampulla ext. and crista ampullaris discharging into it ; r = upper space of the sacculus ; n = nerve bundles of the ramus vestibuli, passing to the utriculus and the macula acust. utric. ; p p″ = adherent wall of the sacculus ; p″ = anterior wall of the sacculus ; n″ = nerve bundles passing to the macul. acust. sacculi ; an = nerve anastomoses running from the bundle n to n″. After a preparation in my collection.

ovalis, to the medial wall of the vestibule, while, between the free lateral side of the saccule and the lateral vestibular wall, there exists a considerable perilymphatic space (Steinbrügge), which is named by Retzius cisterna perilymphatica. To become accurately acquainted with this space, which, below and in front, merges into the scala vestib. of the cochlea, and with the topographical position of the utriculus,

* Lehrbuch der Ohrenheilkunde, 2. Aufl, 1887, S. 459.

both must be examined together on serial sections, made in a frontal direction through the pyramid. Rarely only does one succeed in finding, on horizontal and frontal sections, either the discharge of the ductus endolymphaticus from the medial wall of the vestibule into the two saccules (as represented by Böttcher, l.c. Plate IV.), or the connection between these by the canalis utriculo-saccularis.

The saccules of the vestibule consist of a fibrillar layer of connective tissue, with a structureless homogeneous hyaline membrane, and of an epithelial layer; the former of these layers is most strongly developed near the maculæ acust. (according to Retzius 150—200 μ);

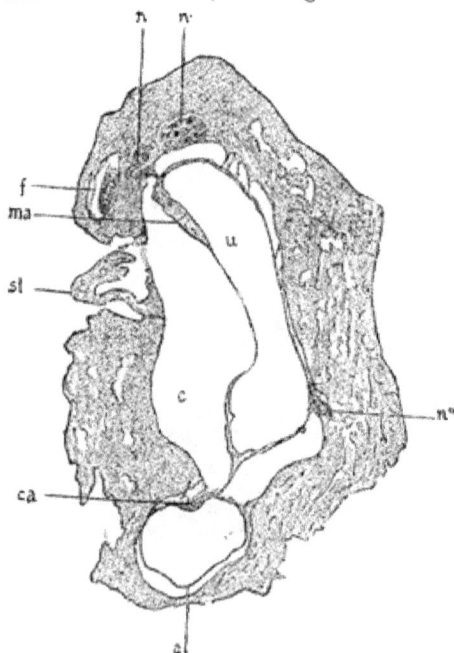

Fig. 154.—Frontal section through the vestibule and the fenestra ovalis of a new-born infant.—u = utriculus, attached to the medial and upper walls of the vestibule; ma = free, lateral wall of the utriculus; st = fenestra ovalis and stapes; c = cisterna perilymphatica; n, n′ n″ = nerve bundles of the ramus vestibuli passing to the utriculus and ampullæ; a = ampulla inferior; ca = crista ampullaris; f = nervus facialis. After a preparation in my collection.

the latter is formed by a simple layer of pavement epithelium. At the maculæ acust. this passes over into a neuro-epithelium (Urban Pritchard), which is made up of the auditory cells, and the filiform cells (Hasse's isolation cells)—(Fig. 155). The auditory cells are bottle-shaped, have a bulging in the middle, and a process at the free surface (Urban Pritchard). The latter formations, which have been termed auditory hairs, consist, according to Retzius, of 10 to 15 filaments which have a length of 20—25 μ. The elliptically-shaped nucleus lies in the bulging portion of the cell. The cylindrical filiform cells have a roundish nucleus placed near the base. The expansion of the ramus vestibuli on the saccules takes place, according to the investigations of Urban Pritchard ('The Termination of the Nerves of the Vestibular and Semicircular Canals,' Quar. for med. Science, 1876),

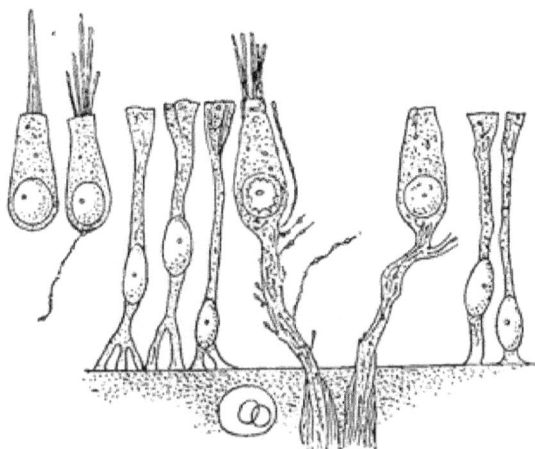

FIG. 155.

in such a manner that the medullated nerve-fibres spread, under numerous anastomoses, in the connective-tissue layer of the maculæ acusticæ, whence isolated non-medullated nerve filaments penetrate the hyaline membrane, pass into the epithelium, and enter into contact with the base of the auditory cells (Retzius). The nerve epithelium of the maculæ acusticæ is covered by a clear, transparent semi-fluid

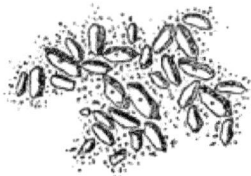

FIG. 156.

substance (Steinbrügge), which after death coagulates, and, as a membrane, envelops the otoliths or otoconia (1—15 μ), consisting of small hexagonal crystals of carbonate of lime (Fig. 156) In the middle of the otoliths Schwalbe found small vacuoles.

2. *Examination of the Ampullæ and Membranous Semicircular Canals.*

The best sections of the ampulla super. et horizontalis are obtained by horizontal cuts through the upper portion of the vestibule (Fig. 153a), but of the ampulla inferior, on frontal sections (Fig. 154a). If the longitudinal cut meet the ampulla in such a way as to expose its embouchure into the utricle and into the semicircular canal (Fig. 153), we first notice a dark stria (a) which crosses the ampulla, and, passing from a slight constriction on the outer surface of the latter, rapidly diminishes towards the middle. This is the point of entrance of the ramus ampullaris, corresponding to which the crista ampullaris arises

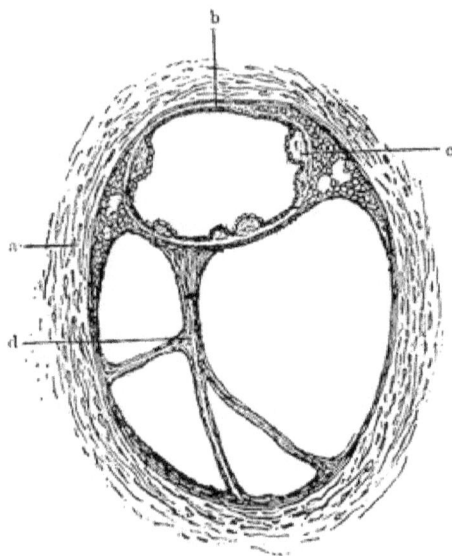

FIG. 157.—Section through the osseous and membranous semicircular canal.—a = osseous semicircular canal ; b = place of attachment of the membranous semicircular canal ; c = elevations on the inner surface of the membranous semicircular canal ; d = vascular bands of connective tissue.

from the inner surface of the ampulla. This crista divides the ampullary space into two unequal portions, the shorter of which, directed towards the utricle, is described by Steifensand as the sinus portion, the longer being known as the tubular portion. The epithelium, which in the ampullæ appears in part flattened, in part cylindrical, shows, on the border of the crista (planum semilun., Steifensand), high cylindrical cells and a neuro-epithelium analogous to that on the maculæ acusticæ of the saccules. Especially beautiful is the appearance of the structure of the crista ampullaris and of its epithelium in transverse sections of the ampulla on preparations treated with hyperosmic acid.

Best adapted for the study of the structure of the membranous semicircular canals, and their relation to the osseous, are transverse sections made perpendicularly to the long axis of the canals. The horizontal and posterior semicircular canals are examined on frontal serial sections, which commence at the posterior extremity of the pyramid, while the superior canal is examined on horizontal sections through the same.

The membranous canals, whose diameter measures about one-third of that of the osseous, are, as shown in the accompanying illustration (Fig. 157), attached with part of their walls to the convex side of the osseous canal (b) (Kölliker, Rüdinger, comp. p. 127), while on the opposite side they are free and bathed by the perilymphatic fluid. From the free part of the membranous canal there pass through its perilymphatic space, which is lined with endothelium, many vascular processes of connective tissue (d) (ligaments) to the periosteum of the osseous canal. Especially numerous are these processes in the new-born infant. They are to be regarded as the residue of the gelatinous connective tissue which fills up the perilymphatic space of the semicircular canals in foetal life. On the inner surface of the canals arise numerous capillary elevations (c) covered with epithelium, which are wanting on the adherent portion of the canal (b), and at the embouchure of the semicircular canals into the utriculus (Rüdinger). Through these the inner superficial space of the canals is considerably increased. They are covered with a polygonal epithelium which, on the band (raphe) lying on the concave side of the canal, and extending into the ampulla, assumes a more cylindrical form. Nerve-elements have not been proved to exist in the semicircular canals.

3. Histological Examination of the Membranous Structures and of the Terminal Apparatus in the Cochlea.

The histological examination of the cochlea is most frequently made on sections carried parallel to the long axis of the modiolus, either in a frontal or in a horizontal direction through the pyramid. Most instructive are those which pass through the modiolus in its entire length from base to apex, because, on them, not only all the convolutions are met in their greatest extent, but also the course of the ramus cochleæ in the modiolus, the disposition of the ganglion spirale, and the entrance of the twigs of the cochlear nerve into the lamin. spiralis, and from here into Corti's organ, are brought most clearly under observation. We ought, however, not to omit to extend our investigations, also to transverse sections directed vertically to the long axis of the modiolus, in order to gain a clear knowledge of the arrangement of the nerve bundles and bloodvessels in its osseous canals, as well as of the relations of these bundles to the spiral ganglion. Since, in sections made parallel to the long axis of the cochlea, Corti's organ is only represented in profile, it should also be examined on surface preparations, these being the only ones on which it is possible to study the dentate projecting borders of the crista spiralis, the arrangement of Corti's pillars, the formation of the lamina reticularis, etc. Especially suitable for this purpose are labyrinths which have been fixed in osmic acid or chloride of gold, and on which, after sufficient hardening in alcohol, the cochlear capsule is broken open, the lamina spiral. ossea et membranacea carefully detached with a needle, and placed upon the glass slip to be examined after the usual methods.

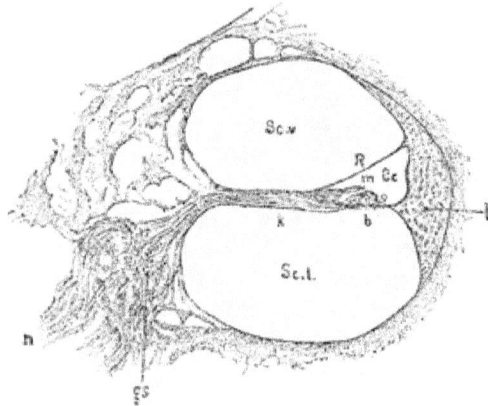

Fig. 158.—Section through the lower convolution of the cochlea of a new-born infant.— Sc. v = scala vestibuli ; Sc. t = scala tymp. ; k = lamina spiral. ossea ; b = lamina basilaris ; l = ligamentum triangulare ; R = membrana Reissneri ; Cc = canalis cochlearis ; o = Corti's organ ; m = Corti's membrane ; n = fasciculus of the ramus cochleæ ; gs = ganglion spirale. After a preparation in my collection.

The study of the details of the complicated cochlear structures must be preceded by a general superficial examination under low magnifying power. A section of one of the coils, enlarged forty to fifty times, is sufficient to see the extent and position of both scalæ (Sc. v., Sc. t.) of the canal. cochlearis (Cc), with the terminal apparatus of the acoustic nerve, and the ganglion spirale (gs). To make the subject clearer, we here give a short description of the section of one of the cochlear convolutions. The cavity of the cochlea (Fig. 158) is divided by the lamina spiralis (k, b) into the scala vestibuli (Sc. v.) and the scala tymp. (Sc. t.). The scala vestib. leads immediately into the vestibule; the scala tymp. is closed at the lowest convolution by the membrana fenestræ rotundæ. The two scalæ are in communication at the apex of the cochlea. The osseous portion of the spiral lamina (k), a plate of bone springing horizontally from the outer surface of the modiolus, consists of two lamellæ connected by vertical bridges, which enclose the nerve bundles, passing from the ganglion spirale (gs). From the free border of the lam. spiral. ossea arises the membranous plate (b), which is attached to the projecting triangular ligam. spirale (l). On the membranous spiral plate three portions are distinguished: the inner, which is perforated by numerous openings for the passage of the fibres of the nerv. cochl. (zona perforata, p); the middle portion, supporting the organ of Corti (zona arcuata); and the outer finely-striated portion (zona pectinata).

The scala vestibuli again is divided by Reissner's membrane (R), passing obliquely from the crista spiralis to the outer wall of the cochlea, into two parts, of which the one, formed by the membrana basilaris, the external cochlear wall (stria vascularis), and Reissner's membrane, is termed the canalis s. duct. cochlearis (Cc). This canal, which is lined with epithelium, contains the terminal apparatus proper, is filled with endolymph, and, communicating below through the canalis reuniens Hensenii with the sacculus, ends in a cul-de-sac at the cupola of the cochlea.

When we have become acquainted with the position of the various structures in the interior of the cochlea by slightly magnifying them, we pass on to the examination of the details of Corti's organ, and of the cochlear membranes, using a higher magnifying power.

The situation of Corti's organ is best demonstrated by profile views of a greater number of microscopic frontal sections. We see (Fig. 159) on such sections, at the upper surface of the outer portion of the osseous spiral plate, a tuberosity (H—crista spiralis, Huschke), which is caused by the thickening of the periosteum, and, with its dentate edge, somewhat resembles a cock's comb. It roofs over a spiral canal (sulcus spiralis int.), lined with cubical epithelium (k). The upper sharp edge of the crist. spiral. is known as the labium vestibulare; the lower, which passes into the membr. basilar., as the labium tympanicum.

Externally from the sulcus spiralis lies Corti's organ proper. This consists of a row of inner (C), and a row of outer (C'), slightly-curved rods (fibres of Corti), which rest with their lower extremities upon the membrana basilar., while their upper ends have a joint-like connection. These two rows of rods form Corti's arch. In the lower angles formed by the rods and the lam. basil. lie two rows of roundish nucleated cells (floor cells).

The rods of the outer row have, at their upper extremities, lamelliform processes, directed outwards, to which is attached a net-like fenestrated membrane, the lamina reticularis (r). It roofs over the outer row of Corti's fibres, and the so-called cells of Corti, or outer hair-cells (äh). These structures (Z) are, in the human subject, arranged in four or five parallel rows (Gottstein), and connected with the terminal fibres of the acusticus through small nerve processes. With their lower attenuated ends (Gottstein's basal process), which, according to Böttcher and Baginsky, do not belong to Deiters's cells, they are fastened to the membrana basilar., while their upper broader extremities, provided with auditory hairs, project through the openings of the lam. reticularis. A row of inner hair-cells (ih) is placed before the sulcus spiral. int., immediately in front of Corti's inner pillar. With Corti's cells are connected those named after Deiters (D), the broader ends of which are directed downwards. Ex-

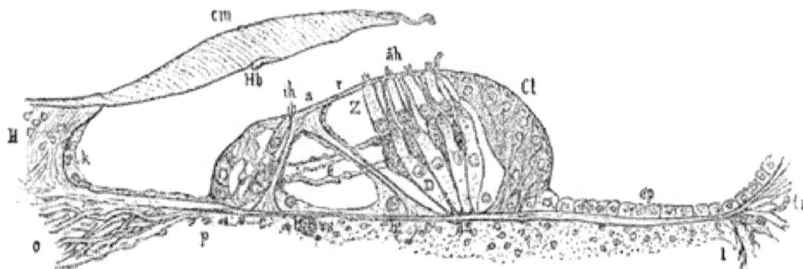

FIG. 159.—Terminal apparatus of the ramus cochleæ, with Corti's organ, in the human subject.—o = lamin. spiral. ossea, with the nerve bundle of the ramus cochl. ; pl = lamin. spir. membr. ; H = tooth of Huschke (crista spiral.) ; c = inner pillar of Corti ; c' = outer pillar of Corti ; r = lamina reticularis ; Z = Corti's cells ; D = Deiters's cells ; ih = inner hair-cell ; äh = four outer hair-cells ; e = radial tunnel fibres of the ramus cochl. passing to the cells of Corti ; k = cells of the sulcus spiralis int.; Cl = Hensen's supporting cells ; cm = Corti's membrane ; vs = vas spirale. After Retzius.

ternally to the last row of Corti's cells lie the supporting cells (Cl) of Hensen, which pass into the epithelium of the outer wall of the ductus cochlearis.

Corti's organ is roofed over by the firm striated membrane (cm) of Corti. It arises near Reissner's membrane from the crista spiralis, and terminates, as is generally supposed, at the outer boundary of the sensory cells of Corti's organ ; it is, according to Böttcher (A. f. O., Bd. xxiv.), attached to the surface of the terminal apparatus of the acusticus.

Reissner's membrane presents a homogeneous structure interspersed with spindle-shaped cells. The side turned towards the canal. cochlearis is covered by a flattened epithelium, the upper side by an endothelium.

The histological examination of the labyrinth in pathological cases, on the importance of which we laid due stress in the introduction to this work, is made according to the methods observed with the normal labyrinth; only the mode of fixation of the tissue-elements differs in a slight degree, since in the examination of pathological labyrinths we mostly use chromic acid and its salts (the fixing fluids of Vlakovic, Urban Pritchard, and Müller's solution, p. 195), rarely only hyperosmic acid and chloride of gold.

By cutting the labyrinth into microscopic serial sections we not only gain information as to the structural changes in its various portions, but, through the successive sections, also a representation of the extent of the morbid process. We employ, by preference, frontal sections made through the pyramid, from the apex of the cochlea to the posterior extremity of the horizontal semicircular canal, because this direction of the cut is more favourable for simultaneous examination of the two labyrinthine fenestræ. Yet there are cases where horizontal sections are preferable, especially when the continuity of pathological products in the vestibule and in the cochlea are to be brought under notice in one section.

The histological dissection of the labyrinth is always to be undertaken where, during life, symptoms of an affection of the auditory nerve had existed, no matter whether it appeared primarily, or complicated with middle-ear disease. It is true the examination often gives a negative result, but much more frequently than is generally supposed, one finds more or less strongly pronounced pathological changes in the labyrinth, which are recognised as the cause of the impairment of hearing during life. We need only refer to the hæmorrhages and purulent exudations in the labyrinth, to the new-formations of connective tissue and bone, to the retrogressive changes in the ganglion cells of the gangl. spirale, in the terminal expansion of the ramus cochleæ in the cochlea, in Corti's organ, and in the macul. acust. of the vestibular saccules.

HISTOLOGICAL EXAMINATION OF THE INTERNAL AUDITORY MEATUS AND OF THE ACOUSTIC NERVE.

The internal meatus and the acoustic nerve are histologically examined on longitudinal and on transverse sections. On longitudinal sections, which pass horizontally through the decalcified pyramid and expose, besides the internal meatus, also the vestibule and cochlea, the direction of the course of the acoustic nerve in the pyramid, the division of the nerve trunk into the posterior upper ramus vestibuli and the anterior lower ramus cochleæ, appear sharply marked (comp. Figs. 100 and 101, p. 127).   The vestibular branch shows, in the neighbourhood of the infundibulum of the internal meatus, a grayish-red swelling, which contains numerous ganglion cells, the intumescentia ganglioformis Scarpæ.   The upper portion of the ramus vestibuli runs in an outward direction in the osseous mass between the cochlea and the vestibule, and, describing a slight curve backwards, enters through the maculæ cribrosæ of the recess. hemiellipt. into the vestibule, to give off small twigs to the recess. utriculi and to the ampullæ of the superior and horizontal semicircular canals.   The lower portion of the ram. vestibuli (the ramus medius of Schwalbe) divides into two small branches, the upper of which also runs forward in the osseous mass between the cochlea and vestibule, enters the latter and proceeds to the neuro-epithelium of the macula acust. sacculi ; while the lower branch passes separately to the crista ampullar. of the inferior ampulla. The bundles of the ramus cochleæ pass through the openings of the tract. spiral. foraminul., in part directly to the first convolution of the cochlea, in part to the nerve-canals of the modiolus, and from here to the lam. spiral. ossea.   Between the fibrous bundles of the modiolus and the lamina spiralis we find, on the periphery of the former, an extensive ganglion layer (zona ganglionaris) interposed (Fig. 158), which lies in the sharply defined canal of Rosenthal, and appears oval in the section (Fig. 160).   Into the lower portion of this ganglion layer the nerve bundles n n′ n″ enter at various spots, and pass, after numerous connections with the ganglion cells at the upper extremity of Rosenthal's canal, into the lam. spiralis.

In the spiral plate, the nerve bundles, which arc through numerous cross anastomoses, in plexiform connection, run with their non - medullated fibres between the two osseous lamellæ to the habenula perforat. of the lower surface of the basilar membrane, pass from here through numerous openings on the upper surface of the same to the ductus cochlearis, and, dividing, according to the investigations of Waldeyer and Gottstein, into very fine, in places knotty terminal fibres, become connected with the inner hair-cells, and, through the gaps of Corti's arch, with Corti's cells, or outer hair-cells.

For the purpose of making transverse sections of the internal meatus, of the acoustic nerve and n. facialis, the pyramid, after having been subjected to one of the usual methods of fixation, and then carefully decalcified, is embedded in celloidin, whereby alone sections of the nerve can be kept *in situ* in the preparation.   Through such serial sections, extending from the entrance of the porus acust.

int. to the floor of the internal meatus, we obtain a correct idea as
to the topographical position of the n. acust. to the n. facialis in the
various portions of the internal meatus, and of the normal relationship
of the transverse sections of the nerve fibres with their sharply defined
axis-cylinders. Such knowledge is all the more important, as it enables
us to judge of the altered appearance of the acusticus in pathological
thickening, connective-tissue proliferation, atrophy, adipose and colloid
degeneration; also of the relations of any neoplasms in the internal
meatus and auditory nerve; of the localization of purulent inflam-

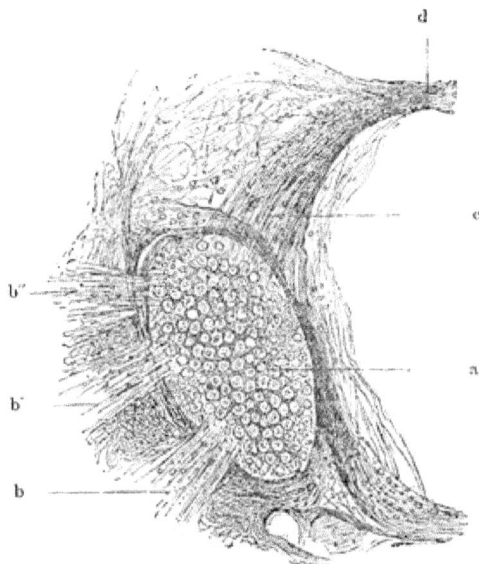

FIG. 160.—Section of Rosenthal's canal and of the ganglion spirale.—a = ganglion cells of
the gangl. spir., lying in Rosenthal's canal ; b b' b" = three separate nerve bundles of the
ram. cochl., entering the gangl. spir. ; c = nerve bundle passing out of the upper pole of
the gangl. spir. ; d = entrance of this bundle into the lam. spir. ossea. After a preparation in my collection.

mations in the internal auditory canal (Gradenigo); of thickenings
and calcifications of its lining membrane; of exostoses, etc.

THE HISTOLOGICAL EXAMINATION OF THE AQUEDUCTS OF THE LABYRINTH.

The histological examination of the aqueducts is not attended with
any special difficulty if the sections be made in a perpendicular direction to the long axis of the canals. For obtaining transverse sections
of the intradural sac of the aquæductus vestibuli, the pyramid, on a
decalcified preparation, with the dura mater in connection, is divided

into two parts by a sagittal longitudinal cut; the inner half is then embedded in celloidin, and the preparation, from the lower portion of the sinus transversus as far as the vicinity of the porus acusticus intern. (p. 132), sliced in serial sections, which are carried nearly vertically to the long axis of the pyramid. The same mode of transverse sectioning is also employed in the examination of the narrow intratemporal part of the aquæductus vestibuli, which, as we have seen (p. 130), describes, in the osseous mass situated in front of the sagittal semicircular canal, a slight curve from the apertura extern. aquæductus vestibuli up to the embouchure into the vestibule. It is only seldom possible to obtain by frontal cuts through the petrous bone a longitudinal section of the entire intratemporal portion of the aquæductus vestibuli.

To make transverse sections of the aquæductus cochleæ on a decalcified pyramid, the funnel-shaped dilatation of the aqueduct is to be looked for at the inner border on the inferior wall (see p. 134), and from here the short canal is divided in a series of sections up to its opening into the scala tympani. On frontal serial sections of the cochlea I succeeded several times in obtaining the entire longitudinal section of the aquæductus cochleæ.

Of the pathological changes in the aqueducts of the labyrinth little is as yet known.

TOPOGRAPHICAL SERIAL SECTIONS OF NORMAL AND PATHOLOGICAL
ORGANS OF HEARING.

When making topographical serial sections of the organ of hearing, it is best to employ, for fixing the tissue-elements, Müller's fluid or Vlakovic's solution (chromic acid and alcohol hardening, p. 195); for decalcification, the solution of nitric acid and chloride of sodium (p. 201); and for embedding, celloidin. As the bulk of the object to be examined is considerable, it should be exposed longer to the action both of the fixing- and decalcifying-fluids than smaller objects. Rapid decalcification especially must be guarded against, on account of the bone nuclei, which it is difficult to get rid of. After decalcification, the preparation is freed from acid by irrigating in a stream of pure water for several hours; it is then placed in dilute alcohol, and only a few hours before embedding in celloidin must it be immersed in absolute alcohol. To allow the celloidin to completely fill up all the cavities of the temporal bone, the tegmen antri mastoid. is broken open, and an aperture filed in the superior semicircular canal. Not until the preparation has been for several days in the thin-fluid celloidin solution should it be transferred to a more concentrated one, and after two to three days more to the syrupy solution, where it is to be embedded.

For the purpose of obtaining topographical serial sections of the auditory organ, it is cut either in a horizontal or a frontal direction by means of the microtome. For the study of the normal relations both methods recommend themselves, as their results supplement one another.

Information as to the topographical representations which are to be obtained from horizontal and frontal serial sections of the ear may be gathered from the previous description of the sections through the temporal bone (p. 33), and from the topographical sections of the organ (p. 147), so that a detailed account may here be dispensed with. It is sufficient to remark that the diameter of the tympanic cavity in the horizontal and frontal planes, the topographical position of the ossicula, the arrangement of their ligaments, and the relative position of the middle ear to the labyrinth and its membranous structures, can by no other method be so accurately and instructively demonstrated as by these serial sections of the entire auditory apparatus. But as this presents in the new-born infant peculiarities differing in some respects from those in the adult, it is recommended, for purposes of comparative study, and for estimating pathological serial sections, to prepare organs of both kinds. The great value of these sections requires, after what has been said, no further proof.

For demonstrating microscopical sections as preparations to be examined with the magnifying lens, a method may here be mentioned which I make use of before large classes. With a sharp knife or a fret-saw two or three square holes are cut side by side in the upper third of the wooden lid of a common cigar-box (22 cm. by 13 cm.). The diameter of these openings measures about 2 cm., so that they are a little larger than an ordinary cover-glass. As they are cut according to the width of the board, there is a distance of 2 cm. between them. The front of the lid is covered with dark-coloured embossed paper, the back with smooth black paper, and in this covering holes are cut which correspond to those in the wood.

The preparation to be examined is now fastened on the board in the following manner : The slide, with its cover-glass directed towards the front, is so placed against the back of the wood that the circumference of the cover corresponds to the square aperture in the board. After this, the slide is pasted over with a piece of black glazed paper which has been gummed and cut out to correspond with the square openings in the board in such a way that the edges of the paper overlap those of the slide. By this means two or three preparations can be fixed side by side on one board, so that, if held against

the light and examined with a lens, they show the exact anatomical
and pathologico-anatomical details in their rough outlines.   In order
to avoid inconvenience while the preparations are being passed from
hand to hand in the class, each board should be provided with a lens
fastened with a piece of cord about a foot long.   A short written
description, or an enlarged drawing of the objects, gummed below the
apertures, will be of assistance.

IX.

HISTOLOGICAL EXAMINATION OF THE CENTRAL COURSE OF THE ACOUSTIC NERVE.

THE histological examination of the medulla oblongata on transverse
sections is, for the study of the topographical relations of the acusticus
and its nuclei, so far of importance to the otologist, as it is only by the
knowledge of the complicated relations of the central course of the
nerve that he will be enabled to judge of the congenital anomalies and
arrest of development in the region of the central origin of this nerve,
and of the primary and secondary pathological changes occurring
therein.

For the investigation of the central origin of the auditory nerve in
the medulla oblongata, of the intra-cerebral course of the fibres of the
acusticus and of the acoustic cortical areas in the temporal lobe, the
chrome-hardening and fixing methods are best adapted.   Most frequently Müller's fluid (kal. bichromici 2.0, natr. sulfurici 1.0, aqu. dest.
100.0) and Erlitzki's solution (kal. bichromici 2.5, cupri sulfurici 0.50,
aqu. dest. 100.0) come into use.

Perfect hardening of these central organs in Müller's fluid is a very
slow process, and often takes weeks and months to accomplish.
Frequent changing of the fluid is the safest means of insuring
homogeneous and rapid hardening of the parts of the brain.   This
is, according to Weigert, materially shortened in the drying apparatus, under a temperature of 30-40° C.   Erlitzki's method effects
more rapid hardening.   To guard against the formation of mould, and
to better preserve the preparations, it is well to add a little camphor
to the above-mentioned fluids.

The examination of the nuclei of the acusticus, and of the course
of its fibres in the central organ, requires special staining methods, of
which especially those recommended by Weigert and Pal have been
found to yield excellent results.*

* Compare Heinrich Obersteiner's excellent work : 'Anleitung beim Studium des Baues
der nervösen Centralorgane im gesunden und kranken Zustande.'—Leipzig und Wien, 1888,
S. 13 u. 15.

1. WEIGERT'S METHOD.

The organs, hardened in Müller's fluid, are, without being washed, placed in alcohol, then in a solution of acetic oxide of copper (two days), then again in alcohol; after that, cutting. Staining in hæmatoxylin 1.0,* alcohol 10.0, aqu. dest. 90, lithium carb. conc. solut. 1.0 two to four hours, until the whole is stained black; decolorization in borax 2.0, ferridcyankal, 2.50, aqu. dest. 200.0. Washing in water, etc.

The medullated fibres appear deep blackish-blue, axis-cylinders, ganglion cells and nuclei bright yellow, almost colourless. Gray substance shows yellowish-white, white substance black.

2. PAL'S METHOD.

A method for the investigation of the central nerve-system, which has of late been frequently employed, and which in some respects differs from Weigert's, is that of Pal, a short description of which is here given.

1. The object is hardened in Müller's fluid, which must be frequently renewed.

2. The hardened preparation is cut under alcohol instead of under water, which is to be avoided.

3. The sections are stained in the following solution: Rp. hæmatoxylin 1, alcohol 10, water 90. The hæmatoxylin solution must be an older one. If fresh, it should have a small quantity of lithium carbonicum added to it.

4. The sections remain for one day in the cold staining medium, and for an hour in one heated to 35-75° C.

5. Washing of the sections in water, and transferring them to a solution of permanganate of potash (0.25 : 100), where they remain until differentiation of the tissues begins to be noticed.

6. Washing of the sections in 1 part kalium sulphurosum, 1 part oxalic acid, 200 parts water, until the brown colour of the preparations has disappeared.

7. Washing in alcohol 1 part, xylol 3 parts, and carbolic acid 3 parts.

8. Clearing up in oil of cloves, and mounting in Canada balsam.

As regards the position of the auditory nuclei and the central course of the n. acusticus, we shall here follow in our description the excellent work of S. Freud, 'Ueber den Ursprung des N. acusticus,' M. f. O., 1886, Nr. 8 u. 9, of which notice was taken in my Lehrbuch, 2 Aufl., 1887.

---

* Rp. hæmatoxylini 1.0, alcohol ab. 100.0, aqu. dest. 100.0, glycerini 2.0, alumin. pulv. 2.0. To be filtered every time it is used. Stains well only after eight days.

A tranverse section through the medulla oblongata in the lowest planes of the acusticus (Fig. 161), through the tip of the great olivary body, shows that this nerve occupies the outer area of the oblongata, which extends to the great ascending root of the trigeminus (V), and contains the section of the restiform body (Cr). Internally and superiorly (dorsal) from the latter lies, beautifully marked against the gray mass, an area (Dk) which has been described as the inner portion of the pedunculus cerebelli, but assigned by Freud to the acusticus, and which is known as Deiters's nucleus. In the outer central corner of the section lies a large nucleus of the auditory nerve, the anterior or external nucleus (8e).

On transverse sections made a little higher, we meet with bundles of the acusticus, which cross the restiform body (VIII$_2$) and have led to

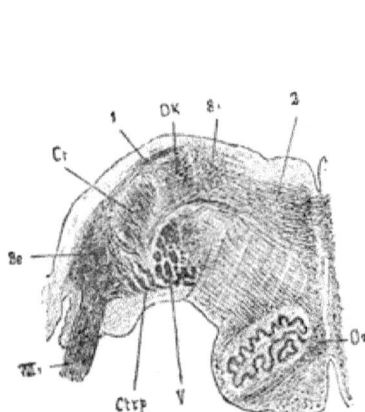

FIG. 161.—Section through the lowest planes of the place of exit of the acoustic nerve, from a human fœtus of 6 lunar months. Treatment with Weigert's hæmatoxylin. —VIII$_1$ = first portion of the acoustic nerve ; 8e = outer, 8i = inner nucleus of the acoustic nerve ; DK = nucleus of Deiters ; V = section of the fifth nerve ; Cr = corpus restiforme ; Oz = interolivary layer ; 1 = acoustic nerve-filaments round the restiform body ; 2 = fibres from 8i to the raphe ; Ctrp = corpus trapezoides.

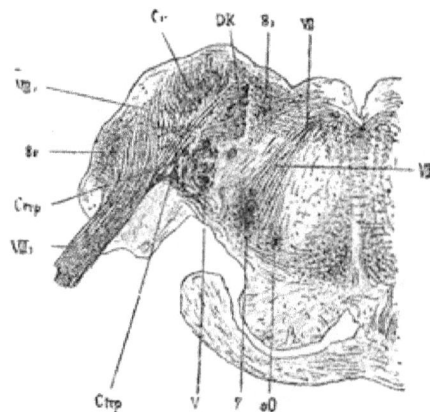

FIG. 162.—Higher preparation from the same series of sections, showing also the nucleus of the facialis 7, and the fibres of the root of this nerve VII, passing to its knee ; VIII$_2$ = the second portion of the acoustic nerve, passing round and through the restiform body ; oO = the superior olive. The rest of the letters as in Fig. 161.

the supposition that the nerve arose out of that fibrous mass. Other portions (VIII$_1$) of the nerve, which occupy the space between the anterior nucleus and the section of the great root of the trigeminus (V), pass over direct into the lateral region of the gray floor (8i). These fibres, the most medial portions of the nerve, do not belong to the acusticus, but to the vestibular nerve, which is connected with it. The planes, in which the course of the vestibular nerve is to be seen, are the lowest ones of the pons. The medulla oblongata here appears connected, through the brachium pontis, with the cerebellum; the roots of the n. facialis (Figs. 163 and 164, VII) and of the n. abducens (Fig. 164) occupy the inner areas of the oblongata. The nucleus of Deiters (Figs. 162 and 163 DK), or the inner portion of the pedunculus cerebelli, can here no longer be demonstrated. The fibres which it contained have curved over into the acusticus (Fig. 163, VIIIa, Roller). The boundary between the tegmental region and that of the pons is given by a large cross bundle, the corpus trapezoides (Figs. 163 and 164, Ctrp), which, while lying free in animals, can also be plainly distinguished in the human subject, although covered over by the fibres of the pons. It arises from the anterior auditory nucleus, is therefore an indirect continuation of the acoustic nerve proper, and passes beyond the middle line. It terminates, at least in part, in a gray

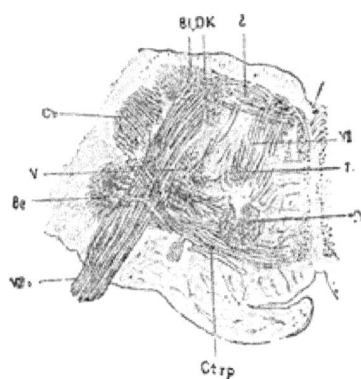

Fig. 163.—Transition of the third portion of the acusticus VIII₃ into the fibres of Deiters's nucleus. Lettering as in Fig. 162.

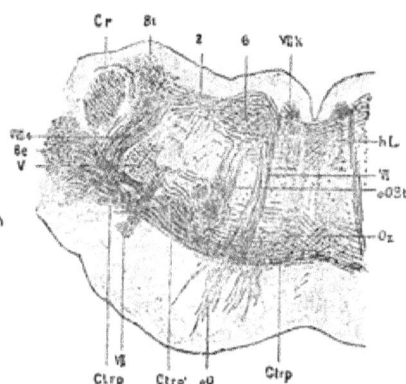

Fig. 164.—Section through the planes of the fourth portion of the acusticus and of the nucleus of the abducens (6).— VI = the n. abducens ; hL = the posterior longitudinal bundles ; VIIk = genu of the facialis ; Ctr′ = portion of the corpus trapezoides, which passes to the superior olive of the same side ; oOSt = peduncle of the superior olive. The other letters are the same as in Fig. 161.

mass lying above it, the superior olivary body (Figs. 163 and 164, oO), which manifests itself, through its connections, as the reflex ganglion of the acusticus (Freud, Bechterew). Other central connections of the acoustic and vestibular nerves pass, as curved fibres (fibræ arcuatæ), beyond the raphe. Among these, special attention has always been directed to the so-called striæ acusticæ, which are visible, with varying distinctness, on the floor of the sinus rhomboideus. They have, however, not the significance of auditory roots, but of central continuations, probably from the anterior nucleus. As to the origin and the central continuations of the n. acusticus, the following is known, according to the investigations of S. Freud.

The n. acusticus (labyrinthine and vestibular nerve) arises from the already described gray substance of the medulla oblongata, and displays the typical relationship of a posterior spinal root. Its central continuations pass mostly beyond the raphe of the oblongata. The outermost and most posterior portion of the nerve (VIII$_1$), the acusticus proper, terminates in the great anterior nucleus (Fig. 161, 8e). Its continuation from here is to be found partly in bundles (Fig. 161), which pass round the restiform body, then backwards towards the raphe, and whose further course is not quite known (striæ acusticæ, connections with the inner nucleus), partly in the large cross bundle of the corpus trapezoides (Ctrp). This mass of fibres is composed of several constituents; one part of them connects, according to Flechsig, the anterior auditory nuclei of both sides with one another, a second bundle runs to the superior olive of the same side (Fig. 164, Ctrp), a larger bundle to the crossed superior olivary body. It is supposed that other bundles pass out of the corpus trapezoides into the interolivary layer, and descend towards the spine (Freud). The superior olivary body appears to be the glanglion for the reflex of auditory impressions on the movements of the eyes, as it is connected with the nucleus of the n. abducens through its so-called peduncle (Fig. 164, ooSt), and with the posterior corpus quadrigemina through the 'lateral loop.' The latter connection was first recognised by Flechsig and Bechterew, and demonstrated by Baginsky through experimental degeneration. The further course of the acusticus, from the posterior corpus quadrigemina into the inner capsule, requires investigation.

A second portion (Fig. 163, VIII$_2$) of the n. acusticus unites with those fibres which pass through and arise from the nucleus of Deiters (DK). This portion is morphologically quite intelligible, as it represents an ascending root (Roller, Freud), such as are also met with in other nerves (trigeminus, vagus). But of its physiological significance nothing is known.

A third portion (Fig. 164, VIII$_4$)—the medial and uppermost fibres—constitutes the nerve of the semicircular canals. It runs partly through the restiform body, partly inwards from the same, and terminates in the inner acoustic nucleus, which, after disappearance of the nucleus of Deiters, occupies its space (large-celled nucleus). Other connections of the vestibular nerve known are : curved fibres to the nucleus of the abducens (Fig. 164, $_2$) and beyond the raphe, occasioning the reflex movements of the eyes in vertigo (the fibres beyond the raphe possibly represent the central continuation in the cerebellum) ; a fasciculus to the interolivary layer, descending towards the spine ; and curved bundles, which pass round the lateral wall of the ventricle, to the crossed central ganglion of the cerebellum (roof-nucleus, spheroidal nucleus, nucleus emboliformis).

Macroscopical preparations of the medulla oblongata for the demonstration of the striæ acusticæ on the floor of the sinus rhomboideus, and of the exit of the trunk of the n. acusticus from the med. obl., are to be placed for several hours in a weak solution of chloride of zinc ($\frac{1}{2}\%$), and then soaked in pure glycerine for ten to fourteen days. The morbidly changed oblongata (tumours, etc.) may be subjected to the same method of preservation. If the preparation be placed in a shallow glass dish and covered with a glass shade in such a way that no dust can penetrate into the enclosed space, the object will retain for years its normal shape and colour.

The microscopical dissection of the medulla oblongata and of the temporal lobe in pathological cases is always advisable whenever the examination on the living subject pointed to a central affection of the acusticus, or where, during life, a high degree of deafness had existed which was not accounted for by the result of the autopsy. Especially with deaf-mutes should the examination of the med. oblong. not be omitted, no matter whether the deafness was congenital or brought on after birth by serious diseases of the middle ear (diphtheritis scarlat.) and of the labyrinth (parotitis), or lastly, by meningitis, cerebro-spinal meningitis and hydrocephalus. In the adult, the examination must be extended to the medulla oblongata and the temporal lobe, if, during life, the disturbance of hearing was complicated with cerebral symptoms, or with aphasia ; also with deafness the result of meningitis, hydrocephalus, apoplexy, and tumours of the brain. But also in deafness of many years' standing, caused by carious-necrotic destruction of the labyrinth, by obliteration and ossification of the labyrinthine cavity, the oblongata must be subjected to microscopical examination, because, as we have seen (p. 264), destruction of the labyrinth in animals (Baginsky) leads to microscopically perceptible secondary changes in the nuclei of the acusticus.

LITERATURE.

Gabrielis Fallopii medici mutinensis Observationes anatomicae.
Coloniae 1562. — *Bartholomäus Eustachius*: Epistola de organis au-
ditus. In eius opusculis anatomicis. Venetiis 1563. — *Hieronimus
Fabricius ab Aquapendente*: Tractatus anatomicus triplex de oculo,
aure et larynge. 1614. — *J. Mery*: Description exacte de l'oreille.
Paris 1677. — *Duverney*: Traité de l'organe de l'ouïe. Paris 1683. —
R. Vieussens: Traité de la structure de l'oreille. Toulouse 1714. —
Duverney: Traité de l'organe de l'ouïe. Leide 1731. — *Joannes
Fridericus Cassebohm*: Tractatus quatuor anatomici de aure humana.
Halae 1737. — *Antonius Maria Valsalva*: Tractatus de aure humana.
Venetiis 1740. — *Dominici Cotunnii*: De aquaeductibus auris
humanae internae. Neapoli 1761. — *Comparetti*: Observationes
anatomicae de aure interna comparata. Patavii 1789. — *Johann
Leonhard Fischer*: Anweisung zur praktischen Zergliederungskunst.
Die Zubereitung der Sinneswerkzeuge und Eingeweide. Leipzig
1793. 8. Cap. IX—XI. S. 81—108. — *Floriano Caldani*: Osser-
vazioni sulla membrana del tympano, etc. Padua 1794. — *Anton
Scarpa*: Anatomische Untersuchungen des Gehörs und des Geruchs.
Nürnberg 1800. — *Joan. Hieronym. Kniephof*: De praeparatione
anatomica organorum auditus. In Act. Acad. nat. curios. Vol. III.
p. 228. — *Samuel Thomas Sömmering*: Abbildungen des menschlichen Gehörorgans. Frankfurt a. M. 1806. — *J. Cunningham
Saunders*: The Anatomy of the Human Ear. London 1806. — *Joh.
Georg Ilg*: Einige anatomische Beobachtungen vom Bau der Schnecke
des menschlichen Gehörorgans. Prag 1821. — *John Shaw*: A
Manual for the Student of Anatomy; containing rules for displaying
the structure of the body, so as to exhibit the elementary views of
anatomy and their application to pathology and surgery. London
1820. 8. edit. 3. 1822. — *David Tod*: The Anatomy and Physiology
of the Organ of Hearing. London 1832. — *M. Gilbert Breschet*:
Etudes anatomiques et physiologiques sur l'organe de l'ouïe et sur
l'audition. Paris 1833. — *Ed. Hagenbach*: Disquisitiones anatomicae

circa musculos auris internae hominis, etc. Basileae 1833. — *E. Alex. Lauth:* Neues Handbuch der praktischen Anatomie oder Beschreibung aller Theile des menschlichen Körpers, mit besonderer Rücksicht auf ihre gegenseitige Lage, nebst Angabe über die Art., dieselbe zu zergliedern und anatomische Präparate zu verfertigen. Band I. Stuttgart 1835. S. 351 u. f. — *Eduard Hagenbach:* Die Paukenhöhle der Säugethiere. Leipzig 1835. — *G. Breschet:* Recherches anatomiques et physiologiques sur l'organe de l'ouïe et sur l'audition dans l'homme et les animaux vertébrés. Paris 1836. — *Arnold:* Icones organorum sensuum. Turici 1839. — *S. Pappenheim:* Die specielle Gewebelehre des Gehörorgans nach Structur, Entwicklung, Krankheit. Breslau 1840. — *Josef Hyrtl:* Vergleichend - anatomische Untersuchungen über das innere Gehörorgan. Prag 1845. — *Corti:* Recherches sur l'organe de l'ouïe des mammifères. Zeitschr. für wissensch. Zool. III. 1851. — *Kölliker:* Mikroskopische Anatomie. II. 2. 1852. — *J. Toynbee:* On the structure of the ear. London 1853. — *Toynbee:* Transactions of the Pathol. Soc. 1853. Vol. IV. — *Joseph Toynbee:* Descriptive catalogue of preparations illustrative of the Diseases of the Ear, p. 121. London 1857. — *J. Gerlach:* Mikroskopische Studien auf dem Gebiete der menschlichen Morphologie. Erlangen 1858. — *Tröltsch:* Die Untersuchung des Gehörorgans an der Leiche. Virch. Arch. Bd. XIII. S. 513. 1858. — *Otto Deiters:* Untersuchungen über die Lamina spiralis membranacea. Bonn 1860. — *A. v. Tröltsch:* Die Anatomie des Ohres. Würzburg 1861. — *R. Voltolini:* Die Zerlegung und Untersuchung des Gehörorgans an der Leiche. Inaug.-Dissert. Breslau 1862. — *V. Hensen:* Studien über das Gehörorgan der Decapoden. Leipzig 1863. — *August Lucae:* Anatomisch-physiologische Beiträge zur Ohrenheilkunde. Virchow's Archiv. Bd. 29. Berlin 1863. — *Carl Bogislaus Reichert:* Beitrag zur feineren Anatomie der Gehörschnecke des Menschen und der Säugethiere. Berlin 1864. — *B. Löwenberg:* Etudes sur les membranes et les canaux du limaçon. Paris 1864. — *Rüdinger:* Ein Beitrag zur Anatomie und Histologie der Tuba Eustachii. Sep.-Abdr. a. d. Aerztl. Intell.-Blatt. Nr. 37. München 1865. — *Ludwig Mayer:* Studien über die Anatomie des Canalis Eustachii. München 1866. — *Rüdinger:* Ueber das häutige Labyrinth im menschlichen Ohre. Aerztl. Intell.-Blatt. München 1866. — *L. Joseph:* Osteologischer Beitrag über das Schläfebein und den in ihm enthaltenen Gehörapparat. Zeitschr. f. rat. Medicin. Bd. 28. 1866. — *E. Zaufal:* Die pathologisch-anatomische Untersuchung der Gehörorgane. Wien. medic. Wochenschrift. Jahrgang 16. Nr. 62—65. 1866. — *M. V. Odenius:* Ueber das Epithel der Maculae acusticae beim Menschen. Sep.-Abdr. a. M. Schultze's Arch. f. mikr. Anat. Bd. III. 1867. —

Gruber: Beiträge zur Anatomie des Trommelfells. Wochenbl. d. Ges. d. Aerzte in Wien. I. 21. 1867. — *Prussak:* Ueber die anatomischen Verhältnisse des Tommelfells zum Hammer. Sep.-Abdr. a. d. Centralblatt f. d. med. Wissensch. 1867. Nr. 15. — *Rüdinger:* Atlas des menschl. Gehörorgans. München 1867. — *Luschka:* Der Schlundkopf des Menschen. Tübingen 1868. — *Arthur Böttcher:* Ueber Entwicklung und Bau des Gehörlabyrinths nach Untersuchungen an Säugethieren. Dorpat 1869. — *Helmholtz:* Die Mechanik der Gehörknöchelchen und des Trommelfells. Pflügers Archiv. I. 1869. — *Kessel:* Nerven und Lymphgefässe des menschlichen Trommelfells. Medic. Centralbl. Nr. 23 u. 24. 1869. — *A. Politzer:* Ueber gestielte Gebilde im Mittelohre des menschlichen Gehörorgans. Vorläufige Mittheilung. Wiener med. Wochenschrift. Nr. 93. 20. Nov. 1869 u. Arch. für Ohrenheilk. Bd. 5. 1870. — *J. Kessel:* Beitrag zum Bau der Paukenhöhlenschleimhaut des Hundes und der Katze. Centralbl. f. die med. Wissensch. 1870. Nr. 6. — *Wendt:* Ueber schlauchförmige Drüsen der Schleimhaut der Paukenhöhle. Leipz. Arch. f. Heilk. Bd. 11. 1870. — *v. Winiwarter:* Untersuchungen über die Gehörschnecke der Säugethiere a. d. 61. Sitzungsber. d. Acad. d. Wissensch. Mai 1870. — *J. Gottstein:* Ueber den feineren Bau und die Entwicklung der Gehörschnecke beim Menschen und den Säugethieren. Dissert. Bonn 1871. — *Magnus:* Ueber die Gestalt des Gehörorgans bei Thieren und Menschen. Virchow-Holtzendorff's Sammlg. Nr. 130. 1871. — *Nuel:* Beitrag zur Kenntniss der Säugethierschnecke. 1871. — *H. Burnett:* Ueber das Vorkommen von Gefässchlingen im Trommelfelle einiger niederer Thiere. Monatsschr. f. Ohrenheilk. Nr. 2. 1872. — *Rüdinger:* Das häutige Labyrinth. Stricker's Handb. d. Lehre von den Geweben. II. 1872. — *Josef Hyrtl:* Die Corrosions-Anatomie und ihre Ergebnisse. Wien 1873. — *Hyrtl:* Lehrb. der Anatomie. XI. Aufl. S. 307. — *A. Politzer:* Zur mikroskopischen Anatomie des Mittelohrs. Arch. f. Ohrenheilk. Bd. 6. 1873. — *Burnett:* Bloodvessels in the Membrana Tympani. Repr. fr. the 'Amer. Journ. of Med. Sciences.' Philad. 1873. — *C. Hasse:* Die vergleichende Morphologie und Histologie des häutigen Gehörorgans der Wirbelthiere. Leipzig 1873. — *A. Politzer:* Zehn Wandtafeln zur Anatomie des Gehörorgans. Wien 1873. — *S. Moos:* Beiträge zur Anatomie und Physiologie der Eustachischen Röhre. Wiesbaden 1874. — *Urbantschitsch:* Beiträge zur Anatomie der Paukenhöhle. Archiv f. Ohrenheilkunde. Bd. 2. 1874. Wiener med. Presse. Bd. 14. 1873. — *Gerlach:* Zur Morphologie der Tuba Eust. Sitz.-Ber. d. phys.-med. Soc. in Erlangen. 8. März 1875. — *Rüdinger:* Beiträge zur Anatomie des Gehörorgans, etc. München 1876. — *Weber-Liel:* Die Membrana tympani secundaria. Monatsschrift f

Ohrenheilk. Nr. 1. 4. 5. 1876. — *E. Zuckerkandl:* Ueber die Vorhofswasserleitung des Menschen. Sep.-Abdr. a. d. Monatsschr. f. Ohrenheilk. Nr. 6. 1876. — *S. Moos:* Die Blutgefässe und der Blutgefässkreislauf des Trommelfells und Hammergriffs. Sep.-Abdr. a. d. Arch. f. Aug.- u. Ohrenheilk. Bd. VI. 1877. — *Ch. H. Burnett:* The Ear, its Anatomy, Physiology, and Diseases. Philadelphia 1877. — *Schalle:* Eine neue Sectionsmethode für die Nasen-, Rachen- und Gehörorgane. Virchow's Archiv. Bd. 71. Berlin 1877. — *A. v. Tröltsch:* Lehrbuch der Ohrenheilkunde. VI. Aufl. Leipzig 1877. — *H. Schwartze:* Pathologische Anatomie des Ohres. Halle-Berlin 1878. — *E. Zuckerkandl:* Zur Anatomie des Warzenfortsatzes. M. f. O. Nr. 4. 1879. — *E. Zuckerkandl:* Ueber die Venen der Retromaxillargrube und deren Beziehung zu dem Gehörorgane. M. f. O. Jahrg. X. Nr. 4. S. 51. — *W. Kirchner:* Ueber das Vorkommen der Fissura mastoideo squamosa und deren praktische Bedeutung. Archiv f. Ohrenheilk. Bd. 14. 1879. — *Kuhn:* Beiträge zur Anatomie des Gehörorgans. Bonn 1880. — *W. Kiesselbach:* Beitrag zur normalen und pathologischen Anatomie des Schläfebeins, mit besonderer Rücksicht auf das kindliche Schläfebein. A. f. O. Bd. XV. 1880. — *Gustav Retzius:* Das Gehörorgan der Wirbelthiere. Morphologisch-histologische Studien. I. Das Gehörorgan d. Fische u. Amphibien. Stockholm 1881. II. Das Gehörorgan der Reptilien, der Vögel und der Säugethiere. 1884. — *G. Sapolini:* Un tredecesimo nervo craniale. Milano 1881. — *Friedrich Bezold:* Die Corrosions-Anatomie des Ohres. München 1882. — *G. J. Wagenhäuser:* Beiträge zur Anatomie des kindlichen Schläfebeins. A. f. O. XIX. 1883. — *Schwabach:* Das Trommelfell am macerirten Schläfebein. Berlin 1885. — *H. Steinbrügge:* Zur Corrosions-Anatomie des Ohres. Centralblatt der med. Wissenschaften. Wien 1885. Nr. 31. — *Th. L. W. v. Bischoff:* Führer bei den Präparirübungen. München 1886. 2. Aufl. Bearb. v. Prof. N. Rüdinger. — *E. Kaufmann:* Ueber ringförmige Leisten in der Cutis des äusseren Gehörgangs. Wiener med. Jahrbücher. 1886. — *G. Schwalbe:* Lehrbuch der Anatomie des Ohres. Erlangen 1887. — *Prof. C. Toldt:* Lehrb. d. Gewebelehre mit besonderer Berücksichtigung des menschl. Körpers. 3 Aufl. 1888. — *Siebenmann:* Ueber die Injection der Knochencanäle des Aquaeductus vestibuli et cochleae mit Wood'schem Metall. Verhandl. der naturforsch. Ges. in Basel. VIII. Theil. 3. Aufl. 1889.

INDEX OF AUTHORS.

THE END.

Baillière, Tindall & Cox, 20 & 21, King William Street, Strand.